Produktion und Logistik

Diese Reihe dient der Veröffentlichung neuer Forschungsergebnisse auf den Gebieten der Produktion und Logistik. Aufgenommen werden vor allem herausragende quantitativ orientierte Dissertationen und Habilitationsschriften. Die Publikationen vermitteln innovative Beiträge zur Lösung praktischer Anwendungsprobleme der Produktion und Logistik unter Einsatz quantitativer Methoden und moderner Informationstechnologie.

Kontakt

Professor Dr. Thomas S. Spengler
Technische Universität Braunschweig
Institut für Automobilwirtschaft und Industrielle Produktion
Mühlenpfordtstraße 23
38106 Braunschweig

Chetan Talwar

Transition Pathways towards a Low Emission Air Transport System of the Future – A System Level Analysis

Springer Gabler

Chetan Talwar
Technische Universität Braunschweig
Braunschweig, Germany

This book is based on the doctoral dissertation titled "Transition Pathways Towards a Low Emission Air Transport System of the Future: A System Level Analysis," submitted to the Technische Universität Braunschweig, Carl-Friedrich-Gauß-Fakultät (Faculty 1), Institute of Automotive Management and Industrial Production (AIP), Mühlenpfordtstraße 23, 38106 Braunschweig, Germany, in fulfillment of the requirements for the degree of Doctor of Economics (Dr. rer. pol.). This research was funded by the Deutsche Forschungsgemeinschaft (DFG, German Research Foundation) under Germany's Excellence Strategy—EXC 2163/1—Sustainable and Energy Efficient Aviation—Project-ID 390881007.

ISSN 1431-6846 ISSN 2628-0108 (electronic)
Produktion und Logistik
ISBN 978-3-658-46506-3 ISBN 978-3-658-46507-0 (eBook)
https://doi.org/10.1007/978-3-658-46507-0

This Springer Gabler imprint is published by the registered company Springer Fachmedien Wiesbaden GmbH, part of Springer Nature.
The registered company address is: Abraham-Lincoln-Str. 46, 65189 Wiesbaden, Germany

Foreword

This thesis was written as part of my work as a research assistant at the Chair of Production and Logistics at the Institute of Automotive Management and Industrial Production at the Technical University of Braunschweig. My journey over the past 4.5 years has been marked by numerous highlights and learning experiences, each contributing to my personal and professional growth. Working on innovative research within the Sustainable and Energy-Efficient Aviation (SE2A) cluster in the Junior Research Group A1 (JRG-A1) has been both challenging and inspiring. I would like to take this opportunity to thank the many individuals who have supported and guided me throughout this period.

First and foremost, my deepest gratitude goes to my doctoral supervisor, Professor Dr. Thomas S. Spengler. His unwavering support, compassion, and insightful guidance have been instrumental in my academic journey. Professor Spengler not only made me feel comfortable in his institute but also provided intuitive insights that significantly advanced my research.

I am profoundly grateful to Dr. Imke Joormann, who has been a pillar of support throughout my research. Her invaluable assistance in honing my writing and presentation skills greatly improved the clarity and precision of my articles and thesis. Special thanks to Professor Dr. Manbir Sodhi for his role as the reviewer of my thesis. His insightful feedback on my work has been greatly useful. I would also like to extend my sincere thanks to Professor Dr. David M. Woisetschläger for chairing the doctoral committee.

Ms. Cosima Meyer deserves special mention for her support with organizational aspects around the institute. Her assistance made my stay in Germany smoother and more enjoyable, contributing to my overall performance and well-being during my PhD. I am grateful to my colleagues, Dr. Christian Weckenborg and Dr. Kerstin Schmidt, for their vital inputs as senior researchers at the institute. Their constructive feedback and guidance have been instrumental in shaping me into a better researcher. To my fellow researchers, Raphael Ginster, Dr. Alexander Barke, Dr. Christian Scheller, Jan-Linus Popien, and Patrick Schumacher, thank you for your collaboration and support in providing information and sources during the writing and publication of my articles. I also extend my gratitude to Tjard Bätge, Patrick Oetjegerdes, Karen Ohmstede, Yannik Graupner, Moritz Proff, Hongfeng Gao, and Haniye Mobarezkhoo for being part of my journey and wish them the best in completing their PhDs.

My appreciation goes to colleagues from the Chair of Services Management: Julian Delius, Antje Fricke, Jan N. Gremmel, Bastian Land, Vanessa O'Neill, Nadine Pieper, and Rick Pingel. Thank you for your support and for being part of my journey.

I am also thankful to my former and external colleagues, Cora Buchenberger, Dr. David Kik, Professor Dr. Christian Thies, and Bastian Vorwerk, for their kindness and assistance. Special thanks to Professor Dr. Karsten Kieckhäfer, whose system dynamics air transport system model was instrumental in my research. His insights greatly aided in extending the model to address my research questions.

On a personal note, I owe my deepest gratitude to my family. To my parents, Brij Mohan Talwar and Usha Talwar, thank you for your unwavering support and for providing me with the opportunity to pursue my studies abroad. To my wife, Ichha Bhan, thank you for being the light and sunshine of my life, for your positivity, and for standing by me through the most challenging times. To my brother, Jatin Talwar, and his wife, Sonja Peschke, thank you for your continuous support and encouragement throughout my journey.

Lastly, I am profoundly grateful to my spiritual master, Dr. Kamlesh D. Patel, for being the guiding light in my spiritual practice. The Heartfulness practice and my fellow practitioners have been central to my personal development, providing me with the wisdom, patience, perseverance, and character essential for completing this thesis. The practice of meditation provided me with the art of contemplation, which has been a source of strength, helping me navigate the complexities of my research with courage and self-belief.

Chetan Talwar

Zusammenfassung[1]

Das Luftverkehrssystem ist einer der wichtigsten Wirtschaftszweige der Welt, dessen globale wirtschaftliche Auswirkungen auf 2,7 Billionen Dollar geschätzt werden und 3,6% des weltweiten Bruttoinlandsprodukts entsprechen. Das jüngste Wachstum des Luftverkehrs hat jedoch zu einer zunehmenden Besorgnis über dessen Auswirkungen auf das Klima geführt. Zu den schädlichsten Treibhausgasemissionen des Luftverkehrs zählen CO_2, NO_X, Ozon (O_3), Kohlenmonoxid (CO) und Kondensstreifen. In Anbetracht der schädlichen Auswirkungen dieser Treibhausgase haben die Regulierungsbehörden Initiativen ergriffen, um die Emissionen der Luftverkehrsbranche im Einklang mit dem 1,5°C-Ziel des Pariser Abkommens zu reduzieren. Eines der wichtigsten Ziele für das Luftverkehrssystem ist das Erreichen der CO_2-Emissionsziele der IATA net zero 2050 und Flightpath 2050. Um diese Ziele zu erreichen, können neue Antriebstechnologien wie Elektroflugzeuge, Flugzeuge mit Wasserstoffantrieb und nachhaltige Flugkraftstoffe (SAFs) eine wichtige Rolle spielen. Die IATA ist der Ansicht, dass fast 65% der für das Erreichen der Netto-Null-Emissionen bis 2050 erforderlichen Emissionen durch SAF erreicht werden können und 13% der Emissionen durch den Einsatz neuer Technologien reduziert werden können. In Anbetracht dieser Entwicklungen gibt es viele Unwägbarkeiten im Hinblick darauf, wie sich der Übergang des Systems in naher Zukunft vollziehen kann, da es angesichts der technologischen Entwicklungen, der politischen Initiativen

[1] To write the Zusammenfassung, DeepL translator from English to German was utilized https://www.deepl.com/translator

und des anhaltenden Wachstums in der Branche schwierig ist, sich die Aussichten für den Übergang vorzustellen. Es kann argumentiert werden, dass die Entscheidungsträger und die Politik ehrgeizige Ziele für die Industrie festgelegt haben, ohne die Details über Produktions- und Lieferverzögerungen, Brennstoff- und Ökostromanforderungen und Unsicherheiten zur Erreichung dieser Ziele zu erläutern. Akademische Autoren haben sich in der Vergangenheit mit Problemen im Zusammenhang mit der Luftfahrtindustrie befasst (Liehr et al. 2001; Sgouridis et al. 2011; Csala and Sgouridis 2013; Pierson and Sterman 2013; Kieckhäfer et al. 2018) Eine Analyse, die Interaktion der Teilsysteme auf einer hohen Ebene berücksichtigt, zusammen mit Überlegungen zu neuen potenziellen Technologien und der Einführung von Politiken, um verschiedene Pfade des Nachhaltigkeitsübergangs zu bewerten, ist jedoch bisher nicht durchgeführt worden. Aus diesem Grund besteht keine Klarheit über die Zukunft des Luftverkehrs im Hinblick auf die Auswirkungen der Einführung neuer Technologien, verschiedener politischer Maßnahmen, Verzögerungen bei der Herstellung von Flugzeugen, Nachfragewachstumsmuster und Verzögerungen bei der Kraftstoffherstellung auf das Luftverkehrssystem aus einer hochdynamischen Sicht. Darüber hinaus hat bisher noch keine Studie die Trennung von Flugsegmenten und den Treibstoff- und Strombedarf auf Systemebene berücksichtigt, um die Wege zur Emissionsreduzierung zu erkunden.

In Anbetracht der Lücke in der Literatur, der neuen Technologien und der Nachhaltigkeitsinitiativen bestand das Ziel dieser Arbeit darin, verschiedene Übergangspfade herauszufinden, die dazu beitragen können, die von den Regulierungsbehörden für die Luftverkehrsbranche festgelegten nachhaltigen Klimaziele zu erreichen. Konkret sollen in der Studie verschiedene Szenarien vorgestellt werden, in denen neue Technologien dazu beitragen können, die Emissionsminderungsziele für verschiedene Flugsegmente zu erreichen, wobei zahlreiche Unsicherheiten und politische Initiativen berücksichtigt werden. Das sekundäre Ziel dieser Arbeit ist es, die Anforderungen an die Auslieferung von Flugzeugen, die Treibstoffproduktion und die politischen Initiativen aufzuzeigen, die für das Erreichen der nachhaltigen Ziele der Luftfahrtindustrie bis 2050 wesentlich sind. Diese Arbeit verfolgt einen zweigleisigen Ansatz, um ihre Ziele zu erreichen. Erstens sollen den Akteuren der Branche Wege zur Erreichung der ehrgeizigen Ziele aufgezeigt werden, indem der potenzielle Beitrag neuer Technologien in verschiedenen Flugbereichen analysiert wird. Dadurch erhalten die Entscheidungsträger und Politiker einen ersten Eindruck von den Möglichkeiten auf Systemebene. Zweitens soll in dieser Arbeit die Lücke in der vorhandenen Literatur geschlossen werden, indem ein Modell entwickelt wird, das Teilsysteme des Luftverkehrssystems wie Fluggesellschaften, Flugreisenachfrage, Hersteller von Flugzeugtreibstoff, Flugzeughersteller, Umwelt und Luftfahrtaufsichtsbehörden einbezieht. Die Entwicklung eines solchen Modells auf Systemebene bietet den Akteuren der Branche und den Entscheidungsträgern die notwendigen Instrumente, um künftige Politiken und Strategien zu testen.

Zur Durchführung der Analyse wurde eine Systemdynamik-Methode verwendet, da sie zum Verständnis komplexer Systeme beiträgt, indem sie deren nichtlineares Verhalten im Zeitverlauf aufzeigt. Diese Methodik hat sich bei der Bewertung verschiedener Strategien und zukunftsorientierter Szenarien auf der Grundlage unterschiedlicher Parameterwerte als äußerst nützlich erwiesen. Das systemdynamische Luftverkehrssystemmodell (ATS) wurde durch die Einbeziehung neuer Technologien wie Elektroflugzeuge, Flugzeuge mit Wasserstoffantrieb und SAFs erweitert. Diese neuen Technologien wurden entsprechend den Berichten und Expertenmeinungen in der Branche zu verschiedenen Zeitpunkten in das Modell aufgenommen. Um das Potenzial der einzelnen Technologien zur Verringerung der Emissionen in den verschiedenen Flugsegmenten zu berücksichtigen, wurde das Modell um Kurz-, Mittel- und Langstreckenflugsegmente erweitert. Außerdem wurde das Modell um ein Segment der Kraftstoffhersteller erweitert, das die Gesamtkapazität der SAF-Produktion, die Wasserstoffproduktionskapazität und die Kapazitäten für die Erzeugung von grünem Wasserstoff mit Ökostrom umfasst. Schließlich wurde das Modell um Mechanismen zum Testen verschiedener politischer Maßnahmen wie Kerosinsteuer, Sitzplatzsteuer, Stromsubvention und Flottenbeschränkungspolitik erweitert. Nach den Erweiterungen wurde das Modell mit den besten systemdynamischen Validierungstechniken validiert und für gültig befunden.

Um die Forschungsfrage zu beantworten, wurde das erweiterte Modell auf zwei Fallstudien angewendet. In der ersten Fallstudie wird untersucht, ob die CO_2-Emissionsziele des Flightpath 2050 durch den Einsatz von Elektroflugzeugen erreicht werden können, wenn politische Maßnahmen, wie z. B. die Besteuerung von Flugzeugtreibstoff, die Subventionierung von Strom, die Besteuerung von Sitzplätzen und die Beschränkung der Flotte, den Einsatz von Elektroflugzeugen verhindern oder fördern. Die Ergebnisse der ersten Fallstudie sollen den politischen Entscheidungsträgern in der EU dabei helfen, fundierte Entscheidungen über den Einsatz von Elektroflugzeugen auf Kurzstreckenflügen mit einer Reichweite von weniger als 1000 km zu treffen. Das Hauptziel der zweiten Fallstudie ist die Analyse der verschiedenen Verbreitungsstrategien für neue Flugzeugantriebstechnologien, nämlich SAF, Elektro-Wasserstoff und Verbrennungs-Wasserstoff in drei verschiedenen Flugreichweitensegmenten. Dazu wurden verschiedene Szenarien unter Berücksichtigung unsicherer Faktoren wie Verzögerungen bei der Herstellung von Wasserstoffflugzeugen und -treibstoffen, Verzögerungen bei der SAF-Herstellung, makroökonomische Schwankungen, Verschiebungen bei der Verteilung der Flugsegmente, Skaleneffekte aufgrund von Produktionssteigerungen und Schwankungen bei der Verfügbarkeit von Strom erstellt.

Die Ergebnisse der ersten Fallstudie deuten darauf hin, dass in den meisten Fällen der Erlass einer Kerosinsteuer, einer Sitzplatzsteuer, einer Flottenbeschränkung und einer Elektrizitätspolitik allein nicht ausreicht, um die CO_2-Emissionsziele des Flightpath 2050 mit Elektroflugzeugen im Kurzstreckensegment zu erreichen. Im Gegenteil, es zeigt sich, dass eine starke Kombination von Maßnahmen erforderlich ist, um das Ziel zu erreichen. Eine starke Kombination bezieht sich hier auf einen höheren Parameterwert der Kerosinsteuer, der Sitzplatzsteuer, der Stromsubvention oder des Flottenbeschränkungsprozentsatzes, der erforderlich ist, um eine höhere Akzeptanz von Elektroflugzeugen zu erreichen. Dies zeigt, dass eine Kombination von Maßnahmen, die den Einsatz von Elektroflugzeugen durch die Einführung von Stromsubventionen fördert, während der Einsatz von konventionellen Flugzeugen durch die Einführung von Flottenbeschränkungen behindert wird, wirksam ist. Allerdings weist die Kombination von Maßnahmen in der Monte-Carlo-Sensitivitätsanalyse ein hohes Maß an Unsicherheit auf. Die Ergebnisse der zweiten Fallstudie zeigen, dass das Erreichen von Netto-Null-Emissionen bis 2050 für die Luftfahrt ein höchst unrealistisches Ziel ist. Der Hauptgrund dafür ist, dass die Anforderungen an die Produktionskapazität für SAF oder Wasserstoff für Wasserstoffantriebskonzepte höchst unrealistisch sind. Vielmehr wurde festgestellt, dass die Anwendung spezifischer Strategien für verschiedene Reichweitensegmente der beste Weg zu sein scheint, die Emissionen im System bis 2050 zu senken. Insbesondere scheint es eine realistischere Strategie zu sein, bei Langstreckenflügen einen Anteil von mehr als 80% SAF am Treibstoffmix anzustreben und im Kurzstreckensegment einen Anteil von 20% oder weniger an wasserstoffbasierten Antriebskonzepten. Dies liegt daran, dass in diesen Szenarien die Erhöhung der Produktionskapazität und andere Ressourcenanforderungen nicht unerreichbar sind.

Insgesamt ergeben sich aus der in dieser Arbeit durchgeführten Analyse eine Reihe von Schlussfolgerungen für das Luftverkehrssystem. Damit das System in der Lage ist, sich bis 2050 in Richtung Netto-Null-Emissionen zu bewegen, müssen die Entscheidungsträger auf eine Kombination von Strategien und Maßnahmen zurückgreifen. Durch die Wahl einer Kombination von Strategien für verschiedene neue potenzielle Technologien in unterschiedlichen Reichweitensegmenten können realistische Ziele für Kraftstoffhersteller, Flugzeughersteller und Fluggesellschaften gesetzt werden. Darüber hinaus zeigen die Ergebnisse im Fall des CO_2-Emissionsziels für den Flightpath 2050, dass eine Kombination von

Strategien ein praktikablerer und effektiverer Weg ist, um Emissionsreduzierungen durch Technologie zu erreichen, als starke einzelne politische Mechanismen. Die Analyse deutet auch darauf hin, dass der Versuch, die Emissionen im Kurzstreckenbereich zu reduzieren, äußerst schwierig ist. Einer der Hauptgründe dafür ist das hohe Flugaufkommen in diesem Segment, der Anstieg der SAF- und Wasserstoffkraftstoffproduktion und die Anforderungen an die Flugzeugproduktion, die höchst unrealistisch sind. Vielmehr sollten sich Politik und Entscheidungsträger mehr auf die Reduzierung der Emissionen von Mittel- und Langstreckenflügen mit SAF konzentrieren, da der Anstieg der Produktionsanforderungen für diese Segmente realistisch erreichbar ist. Was die Technologien betrifft, so ist es für Elektroflugzeuge schwierig, eine wesentliche Rolle bei der Verringerung der Emissionen im Kurzstreckenbereich zu spielen, es sei denn, es werden starke politische Initiativen ergriffen, um die höheren Kosten zu decken. Auch für das Wasserstoff-Brennstoffzellen-Antriebskonzept kann es eine Herausforderung sein, die Emissionen im Kurzstreckenbereich zu reduzieren, wenn keine großen Wasserstoffproduktionskapazitäten vorhanden sind. Mehr noch als die Wasserstoffverbrennungstechnologie können SAF eine entscheidende Rolle bei der Emissionsreduzierung im Mittel- und Langstreckensegment spielen. Um dies zu verwirklichen, müssen die Fluggesellschaften den Kraftstoffherstellern durch Termingeschäfte Anreize bieten. Dies kann kurzfristig zu einer Verringerung der Gewinne der Fluggesellschaften führen, da die Preise für SAF höher sind. Langfristig kann diese Praxis jedoch zu einer Senkung des Preises führen, da die Produktion zunimmt und Größenvorteile erzielt werden. Schließlich ist für die Entwicklung des Systems in Richtung Nachhaltigkeit eine starke gemeinsame Koordinierung der Systemakteure erforderlich, da in dieser Arbeit gezeigt wurde, dass das System sehr dynamisch ist und die von einem Akteur getroffenen Entscheidungen sich stark auf die anderen Akteure im System auswirken.

Contents

Abbreviations

ACARE	Advisory council for aeronautics research in europe
ACI	Association of airports
ACP	Air cargo providers
AI	Artificial intelligence
AIM	Aviation integrated model
ANSP	Air navigation service providers
APU	Auxiliary power unit
ASEAN	Association of southeast asian nations
ASK	Available seat kilometers
ASTM	American society for testing and materials
ATAG	Air transport action group
ATJ	Alcohol to jet
ATM	Air traffic management
ATS	Air transport system
ATSM	Air transport system model
BAU	Business as usual
BECCS	Bioenergy carbon capture and storage
BWB	Blended wing body
CC	Charter carriers
CCUS	Carbon capture utilization and storage
CLD	Causal loop diagrams
CORSIA	Carbon offsetting and reduction scheme for international aviation
DAC	Direct air capture

DLR	Deutsches zentrum für luft- und raumfahrt (German Aerospace Research)
DOC	Direct operating cost
EF	Ecological footprint
EU	ETSEuropean emissions trading scheme
EU	European Union
FSNC	Flag carriers or full service network carriers
FT-SPK	Fischer tropsch synthesis
GAID	Global aviation industry dynamic
GAIDT	Global aviation industry dynamic transition
GDP	Gross domestic product
GHG	Greenhouse gas
HEFA	Hydroprocessed fatty acid esters and fatty acids
HTS	High temperature super conducting electric motors
IATA	International air transport association
ICAO	International civil aviation organization
IEA	International energy agency
LAB	Lithium air batteries
LCC	Low cost carriers
LCO	Lithium cobalt oxide
LFP	Lithium iron phosphate
LIB	Lithium ion batteries
LMO	Lithium manganese
LSB	Lithium sulpur battery
MSE	Mean squared errors
MTOW	Maximum takeoff weight
NCA	Lithium nickel cobalt aluminum oxide
Ni-Cd	Nickel-cadmium
NIS	National innovation system
NMC	Lithium manganese cobalt oxide
OD	Origin destination
ODE	Ordinary differential equations
OWE	Operating weight empty
PAX	Passengers
PEM	Proton exchange membranes

PtL	Power to liquid
RF	Radiative forcing
RPK	Revenue passenger kilometers
SAF	Sustainable aviation fuel
SD	System dynamics
SMR	Steam methane reforming
SOFC	Solid oxide fuel cells
VAT	Value added tax

List of Units

\$/kWh	Dollar per seat kilometer
\$/l	Dollar per liter
US\$	United states dollar
€	Euro
°C	Degree celsius
CO_2 eq/MJ	Carbon dioxide equivalent per mega joule
CO_2-eq	Carbon dioxide equivalent
GWh	Gigawatt per hour
K	Kelvin
Kg	Kilogram
Km	Kilometer
kWh	Kilowatt per hour
kWh/seat*km	Kilowatt hour per seat kilometer
m	Meter
MWh	Megawatt per hour
Nm	Nautical mile
TWh	Terawatt per hour
Wh/kg	Watt hour per kilogram

List of Symbols

–	Negative/Opposite direction movement/Indirectly proportional
||	Delay in transmission of information or material
+	Positive/Same direction movement/directly proportional
CH_4	Methane
CO	Carbon monoxide
CO_2	Carbon dioxide
H_2	Hydrogen
LH_2	Liquid hydrogen
Li/O_2	Lithium oxide
Li/S	Lithium sulpur
LI_2CO_3	Lithium carbonate
LI_3N	Lithium nitride
LIOH	Lithium hydroxide
N	Nitrogen
NO_x	Nitrogen oxide
O_3	Ozone
S	Sulfur

List of Figures

List of Tables

Introduction

1

Transportation is fundamental to human existence, it can be defined as the transfer of people or goods from one location to another. There are many modes of transport such as road, railway, water and air. Out of these modes of transport, the aviation system has become a crucial part of the global economy as it serves as a great connector of cultures, trade, business and is responsible for providing millions of jobs. The air transport industry has exhibited a trend of tremendous growth in the past 10–20 years. The commercial aviation industry has been resilient to economic shocks like the oil crisis of 1978, the terrorist attack of 9/11, the SARS virus crisis in 2003 and the financial crisis in 2008 and has grown more than 2.4 times since 2000 in terms of air traffic (Airbus 2019). In terms of numbers and the impact on the global market, in 2018, the air transport industry transported almost 4 billion passengers resulting in 8 trillion revenue passenger kilometers (RPK). Globally, it has been found that 57% of international tourists travel by air. The industry also contributed to the economy by transporting a total of $8 trillion worth of goods which represented 35% of all international trade. On an overall level, aviation's global economic impact (direct, indirect, induced and tourism catalytic) is estimated at $2.7 trillion, equivalent to 3.6% of world gross domestic product (GDP) (ATAG 2018). Within the European Union (EU), this value accounted for 4.1% of the GDP of all European countries and 26% of the global passenger traffic in 2017. Although the industry had been growing rapidly, the COVID-19 pandemic in 2020 impacted the airline industry deeply and delivered the largest shock to the airline industry since the world war. However, different authors have tried to predict the recovery window for the air transport system in the next few years. Some authors suggest that due to the

C. Talwar, *Transition Pathways towards a Low Emission Air Transport System of the Future – A System Level Analysis*, Produktion und Logistik,
https://doi.org/10.1007/978-3-658-46507-0_1

resilience of the industry and the necessity of air travel the industry might reach pre-COVID-19 levels by mid-2022 in the most optimistic scenario and by 2026 in the most pessimistic scenario (Gudmundsson et al. 2021). In contrast, alternative viewpoints proposed by other authors indicate that pre-COVID-19 levels may be regained as early as 2023 or as late as 2026 (Michelmann et al. 2023).

1.1 Aviation Emissions and Goal Setting

Although the recent growth and development in the aviation industry has had a positive economic influence on tourism, employment trade and business generation, the growth rates of international aviation do not seem to be ecologically sustainable (Goetz and Graham 2004; Chapman 2007; Vespermann and Wald 2011). The recent increases in growth patterns have caught the attention of environmental activists, institutions and scientists due to the harmful environmental effects of the aviation industry. It is found that the aviation industry has been responsible for an impact on climate change due to the emission of greenhouse gases (GHGs), resulting in global warming effects. Amongst the GHGs emitted by aviation fuel combustion, carbon dioxide (CO_2) has been identified as one of the major drivers of global warming. In addition to CO_2 emissions, the combustion of kerosene produces a large number of other harmful GHGs, specifically, indirect greenhouse gases such as nitrogen oxides (NO_x) or carbon monoxide (CO). These gases have no direct greenhouse effect, but they influence the chemical reaction and formation of other greenhouse gases such as ozone (O_3), which are harmful to the earth's atmosphere (NASA 2008; Schmitt and Gollnick 2016). Furthermore, the combustion of kerosene at a higher level of altitude produces water vapor known as contrails. It has been found that contrails promote cloud formation in the sky and as a result of this, the radiation from the sun to the earth is reflected before it leaves the atmosphere, thus increasing the overall warming process (Schmitt and Gollnick 2016). Recent studies indicate that emissions at high altitudes can increase the radiative forcing in the environment by a factor of 3 by 2050 (Bock and Burkhardt 2019). Currently, on an overall global scale, the aviation industry is responsible for 2% of the total CO_2 emissions (ATAG, 2020). Within the EU, this share corresponds to 3.6% (EASA 2019). Even though the numeric magnitude of CO_2 emissions by aviation is lower compared to other sectors, according to the projected industry growth patterns (4–5% annually), their effects can be more drastic in the years to come (Airbus 2019). It is estimated that aviation may be producing as much as 24% of global CO_2 emissions by 2050, compared to roughly 2–3% today (ATAG 2020a). This holds true even after taking into account the effects of COVID-19 on air traffic in 2020, as it is estimated that the passenger demand might reach pre-COVID levels earliest by 2023 and latest by 2026 (Michelmann et al. 2023).

Due to the harmful effects of greenhouse gas emissions on the environment, many climate scientists have suggested that continuation of the same practice over the years could lead to climate change. Specifically, due to the nature of greenhouse gases like CO_2, the overall temperature of the earth can increase as it causes global warming. In the recent few years, a number of emission reducing schemes and pledges have been taken for the air transport industry. The Advisory Council for Aeronautics Research in Europe (ACARE) introduced "Flightpath 2050" (European Commission 2017; ACARE 2020). The program was launched in 2012 with the aim to strengthen the European aviation industry while taking ambitious emission reduction targets into account. The major environmental goals laid down by Flightpath 2050 were to reduce CO_2 emissions by 75%, NO_x emissions by 90% and noise emissions by 65% by the year 2050 compared to the capabilities of new aircraft in the year 2000. In October 2021, the global air transport industry took one further step to resolve to achieve net zero flying by 2050. To make this resolution many organizations and governments played an important part but the most important part was played by the Air Transport Action Group (ATAG), International Civil Aviation Organization (ICAO) and International Air Transport Association (IATA) to outline the strategies to reach the defined goal (ATAG 2021; IATA 2022a). The strategies to reach net zero flying in 2050 consisted of four major aspects namely technological innovation, sustainable aviation fuels (SAFs), operational improvements and market based measures. Furthermore, many other measures in the form of taxes, quotas and emissions trading schemes were discussed. These included taxes like green tax, carbon tax, jet fuel tax, seat tax, air passenger tax and value added taxes. In the market based measure, the quotas and obligations were also included. For instance, the quota for biofuels and the ReFuelEU Aviation initiative taken in Europe under the "Fit for 55" plan. In these initiatives, the aviation system would systematically increase the use of SAFs over the years reaching almost 65% by 2050 (European Commission 2021).

It can be implied from the above discussion that the climate goals set for the industry by the regulators, organizations and action groups cannot be reached with the current technologies and capabilities. To reach the above goal, substantial research in new aviation technologies and policy measures is necessary. In terms of technology, there have been recent developments in the field of sustainable technologies that can potentially help the industry reach the net zero goals by

2050. One of the possibilities that recently has caught attention is the electric aircraft. A fully electric aircraft would rely on batteries for the electrical energy source as the means to power the propulsion system. These types of aircraft are still in the conceptual stage but if they were to become a reality they would be available for short-haul operation only. Furthermore, it is assumed that the electric aircraft will not have any in-flight emissions and would be considered 100% emissions free. Another development has been in the field of sustainable aviation fuels (SAFs). Sustainable aviation fuels (SAFs) are the alternative drop-in jet fuels that can be produced from different renewable feedstocks (Cabrera and Melo de Sousa 2022). SAFs can be used in place of conventional jet fuel and do not require any structural changes to the aircraft or airport fueling infrastructure. In this regard, multiple methods have been found to produce SAFs. Specifically, the major ones include Fischer Tropsch synthesis (FT-SPK), hydroprocessed fatty acid esters and fatty acids (HEFA), alcohol to jet (ATJ) and power to liquid (PtL) (Zhang et al. 2020; Tanzil et al. 2021; IATA 2022b; Cabrera and Melo de Sousa 2022; Hsu et al. 2022). Hydrogen based propulsion technologies have also gained recent attention in the industry. This is primarily due to its significant potential for reducing emissions in comparison to traditional kerosene-based propulsion. Specifically, airbus and NASA have done research into cryoplanes in the past that can have hydrogen as fuel in aircraft with a tank in the fuselage (Verstraete 2013; Baroutaji et al. 2019). Furthermore, recently Airbus and ZeroAvia have pledged to build a hydrogen propulsion plane for civil use by 2035 and 2024 respectively (Airbus 2021; Nicolay et al. 2021).

1.2 Research Objectives

In the context of the discussions done above related to the recent technological development, goals set for the aviation system by authorities, taxation and quota policies and the persistent growth, it is unclear how would the system transition towards a more sustainable future. To explain it a bit further, in the aviation industry currently a lot of planning, road maps and aims have been set for the future but how this planning in terms of system level interaction, uncertainties and delays would play out is highly unclear. Moreover, it appears that the industry professionals and regulators have missed specific details about the requirements necessary for the ambitious transition to a sustainable air transport of the future. From an academic view point, many authors in the past have approached questions related to the air transport industry. Specifically, the topics that have been dealt in the literature in the past have been related to cyclical dynamics in the airline industry, analysis related to carbon pricing and airline industry competition strategy (Liehr et al. 2001; Sgouridis et al. 2011; Csala and Sgouridis 2013; Pierson and Sterman 2013; Kieckhäfer et al. 2018). But an analysis that considers the sub-systems interaction on a high level along with considerations of new potential technologies, introduction of policies to assess different sustainability transition pathways has not been done yet. Specifically, the previous research has not included the adoption of new technologies, the introduction of policies, aircraft manufacturing delays, demand growth patterns, fuel manufacturing delays, flight segment segregation, fuel and electricity requirements together on a system level setting in an attempt to explore the emissions reduction pathways to achieve net zero goals.

To bridge the gap of the literature and provide clarity to the stakeholders of the aviation industry the primary research question that is set for this thesis is: What are the different transition pathways through which the air transport system can transition towards a low emissions and sustainable system in the future? Furthermore, the secondary research question set for this thesis is that if there are transition pathways that can achieve the goals set for the aviation industry then how would these goals be achieved and what would be the requirements for the same?

To this end, Fig. 1.1 shows the objectives and structure of this thesis.

In Chap. 2, the main stakeholders of the air transport system are identified along with their potential roles for achieving low emissions in the near future. The aviation system on a high-level is discussed in Sect. 2.1. In Sect. 2.2, the air travel demand and the aspects associated with the changes in demand are discussed. Sect. 2.3 consists of a discussion on the different types of airlines and fleet planning strategies used by the airline industry. In Sect. 2.4, different types of airports along with their economic importance is explained. The role of regulators and the emission reduction goals set for the aviation industry are discussed in Sect. 2.5. The discussions regarding sustainable aviation fuels and hydrogen as a fuel along with the feedstock limitations are done in Sect. 2.6. In Sect. 2.7, the sub-section of aircraft manufacturers is discussed along with the new technologies that can be adopted to reduce emissions from the system. Lastly, in Sect. 2.8 the requirements for analysis in terms of methodological design and input parameter to conduct the thesis are discussed.

In Sect. 3 the relevant literature related to the research question is evaluated. In Sect. 3.1, the methodology of choosing the relevant literature is explained. In Sect. 3.2, the literature related to the system dynamics and air transport system is discussed. The literature related to usage of system dynamics in policy analysis is done in Sect. 3.3. Lastly, the research done on a holistic level to achieve sustainability in the air transport system is reviewed in Sect. 3.4.

In Chap. 4, the extensions done in the existing air transport system model are explained in more detail. To this end, Sect. 4.1 contains the reasons for extending the model along with the aimed outcome of the extended model. In Sect. 4.2, the extension of the model regarding causal loop diagrams, stock and flow diagrams, and equation formation are discussed. Lastly, in Sect. 4.3 the developed model is validated and tested to generate confidence.

Chap. 5 contains the application of the extended model for two case studies. In Sect. 5.1 the model is applied to the problem of reaching Flightpath 2050 CO_2 emission goals using electric aircraft and different policies. In Sect. 5.2, the model is applied to assess the transition pathways using SAF hydrogen, electric and hydrogen combustion technology in the global ATS model.

In Chap. 6, the implication of the results from the case studies are discussed. Sect. 6.1 consists of the implications of the results for aircraft manufacturers. In Sects. 6.2, 6.3 and 6.4 the implications of the results for fuel producers, regulators, policy and decision-makers and airline industry are discussed respectively. Finally, in Chap. 7 and 8, the conclusions and the summary of the thesis are explained.

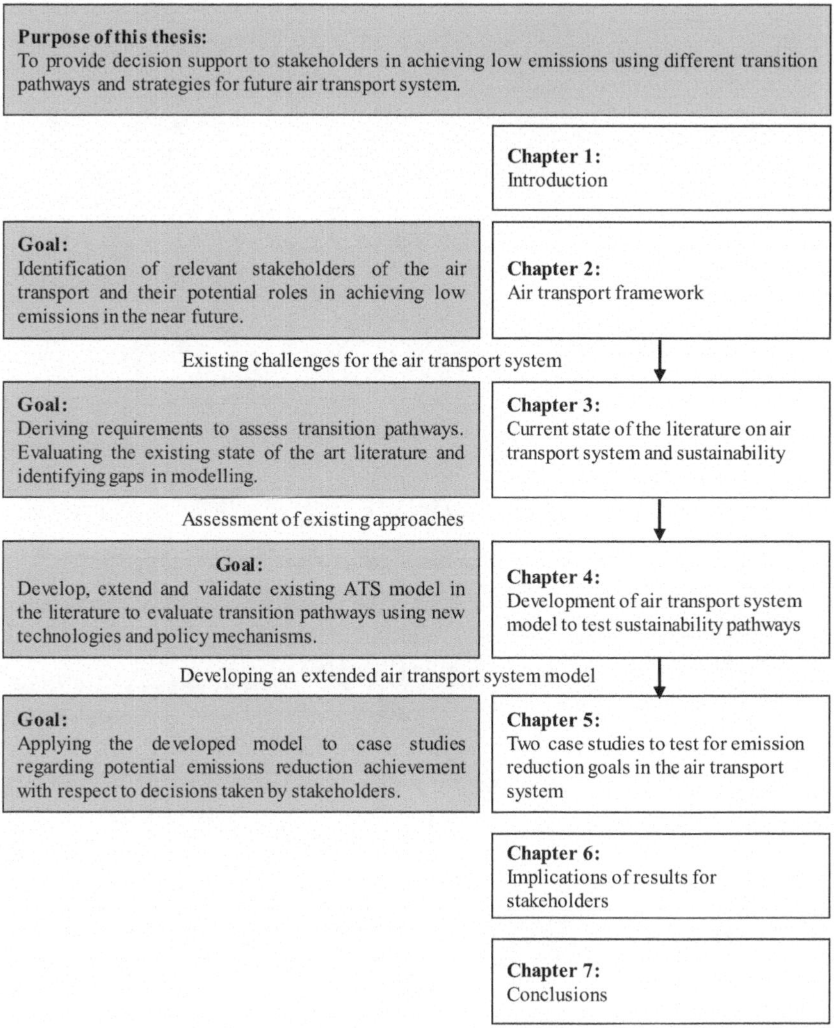

Fig. 1.1 Objectives and structure of thesis

Air Transport Framework 2

In this section, the air transport system and its different components or sub-sections are explained in detail. These include stakeholders like air travel demand, airline companies, airports, environment, aviation regulating authorities, fuel producers and aircraft manufacturers. The aim of this section is to discuss in detail the functioning of each part of the system. Specifically, in this section, the factors that influence the sub-sections are discussed.

2.1 Air Transport System Overview

The air transport system can be defined as a complex system consisting of many sub-systems and elements that interact with each other to form a system that helps in travel and mobility by air. Some of the most important sub-systems of the aviation system are the aircraft manufacturers, airline businesses, air travel customers, airports, aircraft fuel producers, environment and aviation regulatory authorities. Figure 2.1 shows the interaction between the stakeholders in a visual format.

C. Talwar, *Transition Pathways towards a Low Emission Air Transport System of the Future – A System Level Analysis*, Produktion und Logistik,
https://doi.org/10.1007/978-3-658-46507-0_2

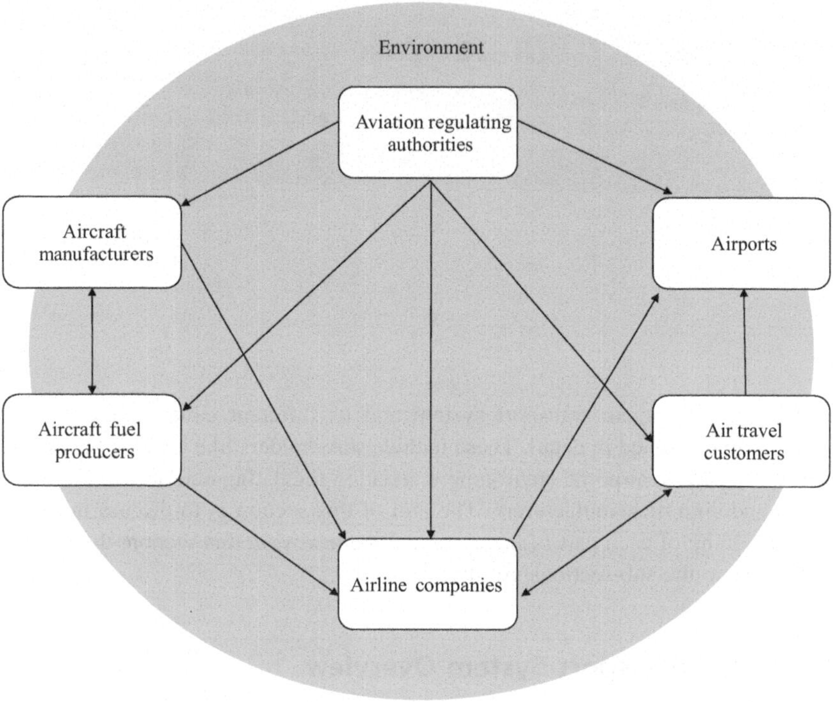

Fig. 2.1 Air transport system overview

The functionality of the air transport system can be explained by explaining the function of each sub-system and its interaction with others in the overall system setting. The aircraft that are developed and produced by the aircraft manufacturers become the products for the airline companies. The airline companies buy the aircraft and operate it. The airline companies assess the air travel demand and based on the trends provide their services to transfer passengers or cargo from one location to another. The airlines distinguish themselves by providing services on different routes, customer services and user experience giving rise to competition in the industry where each type of airline competes with other companies in the same business segment. The fierce competition to gain more market share between the companies sets the airline fares that are available to choose from in the open market for air travel customers. The service routes, seats, services and travel experience offered by the airline companies become the product for the air travel customer. The air travel customers use these services to travel or to ship items across borders which together forms an air travel demand. The airports provide the infrastructure required to carry out the operation of airline companies. The airports are the connecting link between the air travel customers and the airline operators. The airport authorities and staff ensure safety and smooth operations at the ground level for airline companies and air travel customers. The aircraft fuel providers produce different types of fuels required for the different types of aircraft and transport them via road, water or air to the airport. The fuel provided by the fuel manufacturers becomes the product for the airlines which buy the fuel readily available at the airport to conduct their operations. Another aspect that is extremely important for the air transport system is the environment. The environment can be defined as the natural or geographical region in which all living beings operate and conduct economic activity. Aircraft operations to carry passengers and cargo negatively affect the environment as aircraft operations result in greenhouse gases (GHG) that over time increase the temperatures of the earth resulting in global warming. Also, the aircraft fuel producers also affect the environment as they generate GHG emissions to produce the fuel. Lastly, another very important aspect of the air transport system is the regulatory authorities. The regulatory authorities can be defined as organizations, coalitions, cartels, action groups or research institutions that govern the interaction and functioning of airlines, fuel producers, aircraft manufacturers, airports and air travel customers. Regulatory authorities play a very important role in the smooth functioning of the air transport system as they lay down rules, oversee operations, provide new insights and set the safety standards of the aircraft for safe travels by air. The interaction between the sub-systems explained above forms the complex system of air transport. Furthermore, the various interactions between sub-sections give rise to the exchange of information and decisions which ultimately form industry behaviors. Over time, certain industry behaviors dominate the system, the dominating behaviors give rise to a particular type of outcome in the system which in other words can be called industry trends.

2.2 Air Travel Demand

In this section, the air travel demand sub-section of the air transport system is explained in more detail. First, an overview of the factors that influence the air travel demand is provided. Later, the relationship of each influencing factor with air travel demand is discussed.

2.2.1 Air Travel Demand and Influencing Factors

Air travel demand is one of the most important components of the system as without the demand to travel by air the system would not exist in the first place. It is through the desire of the customers to travel by air that the services and products provided by airline companies, aircraft manufacturers, airports and fuel producers are put to good use. Air travel demand in simple words can be defined as the demand by the customers to travel from one point to another by air. It can also be said that the customers desiring to travel by air are willing to forego the other modes of transport like bus, rail and road. It is implicit that the customers are able to derive some extra marginal utility by traveling by air rather than other modes. To explain this, researchers in the past have tried to find the factors that influence air travel demand. In this field, extensive research has been done on different continents using data analysis and regression techniques to find out the correlations between the different factors and the air travel demand. For instance, in the study by (Boeing 2016) the economic drivers like gross domestic product (GDP), labour market, income and trade along with ease of travelling were taken into account. Even though the report was insightful, the detailed aspects of the forecasting methodology were not clearly explained. In a study by (Airbus 2016), the GDP and market maturity aspects i.e., developing vs. advanced economies were assessed as the two main influencers for the demand, but similar to the Boeing study, insights on the methodology used for forecast were not provided. Some other companies like (Intervistas 2007) also prepared a report for the IATA on the finding the demand elasticities of air travel. In the report, they analyzed the effects of changes in income and airline fares on the demand elasticities. It can be argued that many other factors that influence the air travel demand are not discussed in the reports mentioned above. For instance, the economy does not only consist of income and production but also the growth and demography of the population that could affect the demand. Also, nowadays many countries are clustered in cities and the short air travel demand could also be affected by the city clusters. Lastly, activities like business and level of education can be some aspects that could potentially create air travel demand. From this discussion, it can be inferred that factors that affect the air travel demand are difficult to explain in the form of a list. The main reason for this is that studies that explore the factors that affect the air travel demand are fragmented to cater for different geographical regions, population, income and leisure habits. Furthermore, there are a large number of components that affect the air travel demand indirectly and can also be discussed when discussing the factors that affect the air travel demand. The aim of this section is to discuss the most important factors that affect the air travel demand from a systems perspective. For this, the list developed by (Tillmann et al. 2023) is utilized and from the list only the most important factors that affect the demand directly are chosen for discussion.

Figure 2.2 below graphically shows the relationship between the different factors that affect the air travel demand as discussed above from a systems perspective. The boxes in the figure represent the different factors, the air travel demand is shown in the middle. The arrow shows the direction of the relationship thus acting as connectors between the factors and the air travel demand. In the diagram, the "+" symbol signifies a parallel movement between the components, indicating a direct relationship, whereas the "–" symbol denotes an inverse relationship, indicating movement in opposite directions between components. It can be seen in the figure that population, GDP per capita and market demand structures affect the air travel demand positively. In other words, if any of these aspect increase over time then the air travel demand would also increase. On the other hand, air fares affect the demand negatively as an increase in air fares reduces the demand and vice versa. The detailed discussion about each aspect is given as follows:

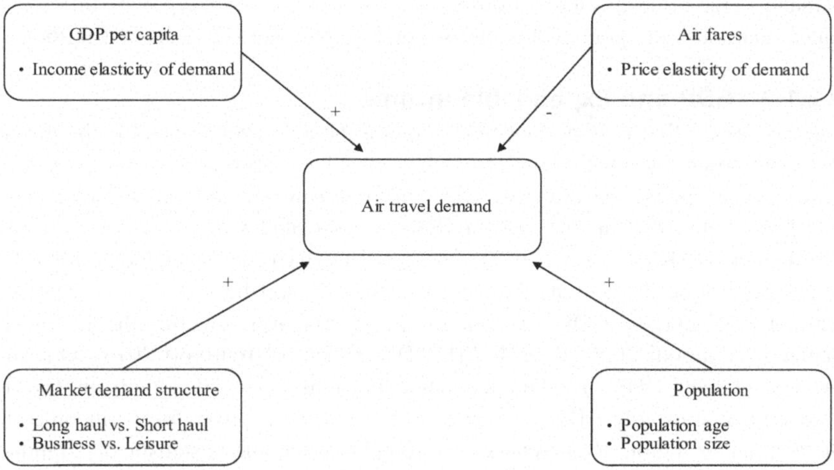

Fig. 2.2 Factors affecting air travel demand

2.2.1.1 Airline Fares

Numerous studies in the past have attempted to examine the impact of airline fares on air travel demand. While there can be many methodologies to do so, price elasticity is one concept that has helped researchers to provide deeper insights into the effects of changes in price on air travel demand. The price elasticity of air travel can be defined as the ratio of the percentage of air travel demand to the percentage change in airfare. There are many nuances to testing the effect of price on air travel demand for instance, the price elasticity of a good directly depends on the possibilities of substitution of that good. A high amount of substitutes will imply a high price elasticity, whereas a low amount of substitutes would result in a rigid demand and low elasticity (Brons et al. 2002). In the case of air travel demand, the effect of price on demand is generally noted to be negative or it is considered to be elastic. In other words, a small increase in airfare would result in large decreases in the demand for air travel and if there are small decreases in the air fares then the demand for air travel increases largely. In the report by (Intervistas 2007), −0.8 was found to be price elasticity for the long haul flights in the North American region. Although, this helps in understanding the basic aspects of air travel demand it misses out on many other aspects that affect the demand as the price elasticity of different customers and flight types are different. Since that is not the goal of this thesis therefore they are not discussed further.

2.2.1.2 GDP and Expendable Income

Another aspect that is important in explaining the air travel demand is the gross domestic product (GDP). Many research studies have been done in the past that have tried to assess the effect of GDP on air travel demand and it has generally been found that an increase in GDP is associated with an increase in air travel demand (Kluge et al. 2017). These studies have been conducted for different geographical locations like Sweden, the UK and Italy using the regional and national data and GDP was found to be significantly affecting the air travel demand (Andreoni 2006; Kopsch 2012; Department of transport 2013). One of the reason why GDP or GDP per capita is highly associated with air travel demand is that usually GDP per capita has been taken as proxy for the income of the country or region. Researchers have found that income is statistically significant when associated with the air travel demand (Kluge et al. 2017). One simple reason for this relationship could be that an increase in income also increases the disposable income of individuals. An increase in disposable income would mean that people of the region or nation have more to spend on. Therefore, a part of spending could be travelling and air travel could arguably be a better option as air travel is faster than other modes. To prove the association between income and air travel demand, it has been found by IATA that an increase in income has been to a large extent a reason for the growth in air traffic over the past few years (IATA 2020). It is also believed in the literature that the income elasticity of air travel is an important measure of understanding the air travel demand. There are a number of reasons for deeper understanding of the income elasticity but some important ones are; first, the income elasticity is important for demand forecasting purposes as the higher value of elasticity means that greater positive change in the air travel demand. This means that the system operators like airports, airlines and fuel producers can be prepared for the changes. Second, the level of income elasticity helps airline companies in understanding the changes in the way customers view air travel. For instance, it can be viewed either as a luxury (i.e., income elasticity exceeds 1) or a necessity (i.e., income elasticity lies between 0 and 1). Furthermore, this information can be different for different range segments for example, long distance travel can be viewed as a luxury and short distance can be viewed in some cases as a necessity and airlines can use this information to target their strategies accordingly. Third, many studies use income elasticity as the measure of the maturity level of the market (Gallet and Doucouliagos 2014). For instance, an income elasticity exceeding 1 means that the market is immature and has growth potential. An income elasticity of 1 or below means that the market is fully mature or has low growth potential. An income elasticity of 0 means that the market is saturated and has no growth potential (Graham 2000).

2.2.1.3 Long-haul vs. Short-haul Elasticities

Another very important aspect in understanding the influences of air travel demand is the distance. Since from the above discussion, it is already known that price elasticities exist in air travel, it has been found that short-haul travel has high price elasticities as compared to long-haul travel. In other words, passengers traveling abroad or internationally are less averse to price changes than people travelling short-haul distances (Intervistas 2007). There could be many reasons to explain this behavior, but one of the most important reasons for this is the opportunity for intermodal substitution on short-haul air travel. In other words, short-haul air travel can be replaced by car, train or bus if the price of air travel increases. On the other hand, the ability to substitute long-haul air travel is extremely hard and often more expensive. Another reason for the higher price elasticities in short-haul travel could be seasonal travel. In air travel, often it has been found that the demand does not remain the same across the year. The demand usually increases during the summer and winter breaks where most airline companies try to gain the most profits by altering their price strategies. This results in large price differences and since most of the travelers during that period are price sensitive, therefore, this creates higher price elasticity for short-haul air travel.

2.2.1.4 Business vs. Leisure Travelers

It has been found in the literature that there is a difference in price sensitivity based on the type of traveler. The research suggests that all else being equal, business travelers are less sensitive to price changes than leisure travelers (Intervistas 2007). One of the reasons for this discrepancy could be that generally business professionals have less flexibility to delay or cancelations than leisure travelers. On the other hand, leisure travelers are always trying to find the best deals regardless of the dates and times they would travel (Intervistas 2007). Even though business passengers are highly inelastic recently there is a new trend in which the passenger combines both business and private leisure trips. These types of trips can also change the outlook of passengers towards air travel for the outward travel there could be less elasticity but later on for the leisure trip, the passenger could be highly price sensitive. This type of behavior could also influence the demand (Kluge et al. 2017).

2.2.1.5 Population

Apart from the type, distance and the income of the traveler, population is another factor that influences the air travel demand. In their research for IATA, (Intervistas 2007) found that an increase in overall population would increase air travel as the market potential of the customers increases. For example, the population in some Asian countries is far greater than the ones in European countries, therefore, the demand and air traffic is higher in that region. Furthermore, statistics show that the six largest countries of the European region i.e., Germany, France, Italy, Spain, Poland and the United Kingdom make up 70% of the population and have the highest share of air traffic. In terms of the population, it has been found that there are differences in the demand for air travel between the different age groups. It has been found in the research study by (Möller et al. 2007) that in Austria, the age group 55–64 had the most number of travels per year. The travel volume for other group of 65 to 79 and 80 and above was decreasing over the years, some of the reasons being financial stability and health. It was also found that elderly people prefer to travel during off-peak seasons and for a longer time period. Regarding travel by young people, (World Tourism Organization (UNWTO) 2016) conducted a study on people travelling between the age of 15–29 as they accounted for almost 23% of global travelers. The major reason for traveling for this type of group is to learn and experience different cultures.

Furthermore, these types of travelers have low budgets and are highly price sensitive. Also, the idea of travelling abroad for studies has increased the air travel demand from this age group over the past few years.

2.3 Airline Industry

This section is aimed at explaining the airline industry as a sub-section of the air transport system. The section begins by explaining the different types of airline businesses that exists. After this, the general fleet planning strategies adopted by airlines to achieve smooth operations are discussed.

2.3.1 Types of Airline Companies

Airlines industry is an important part of the air transport system as the major operational aspect of the system are carried out by the airline companies. Airline industry consists of airline companies that offer their services to transport passengers and cargo over long distances. The airline flight services are mainly divided into three types of flight segments namely short, medium and long-haul. The different type of airlines provides their transport services in one, two or all three segments based on their business model. It is known in the industry that the companies often make use of the demand or potential demand and existing competition in the segments to choose service offering. Airlines use the different aircraft manufactured by the aircraft manufacturers to successfully transport passengers or cargo. But the type of service and aircraft depends on the type of business of the airline company. In principle, there are four types of airline companies in the world:

1. National or flag carriers or full service network carriers (FSNC)
2. Low cost carriers (LCC)
3. Çharter carriers (CC)
4. Air cargo providers (ACP)

National or flag carriers otherwise known as full service network carriers (FSNC) are the airline companies that offer their passenger transport services in all three flight segments i.e., short, medium and long-haul. These type of companies are called National carriers because most of the countries in the world have a national airline which is either party or completely state owned. The Low Cost Carriers

(LCC) are the airline that have a business model of cost reduction and gain market size by lowering the price. Unlike the FSNCs, the LCC airlines offer their services only in the short to medium range of 500–2000 km range segment. This mainly include short-haul or sometimes medium-haul flights. Sometimes the LCC airlines offer medium-haul flights as the feeder flights for the FSNC networks. Charter carriers (CC) are the type of airlines that offer air transport services for passengers and goods on seasonal basis or occasional demand. According to the international air law they are called "non-scheduled traffic" because these type of carriers do not have a specific type of flight schedule pre-planned (Schmitt and Gollnick 2016). Lastly, air cargo providers (ACP) are the companies in the aviation sphere that offer the services of transporting the goods by air.

2.3.2 Airline Fleet Planning and Strategies

Another very important aspect of the airline industry is fleet planning and strategies. Fleet planning can be defined as actions taken by airline companies to organize, optimize and arrange their aircraft fleet to serve the passenger demand. This process is extremely complex and requires accuracy. This process is usually done in advance before the operation of flights. Furthermore, due to the complexity of the process airline companies either outsource this task to other consultancies or use high level technologies like AI, machine learning or big data to help them for the same. Fleet planning is considered under the operational planning among the airline's business segments. Fleet planning is one task that if done incorrectly can result in bad service and hefty fines for not following the regulations. It can also lead to a loss in customer satisfaction, loss in customer perception and retention and consequently loss in revenue or profits. Generally, it has been found that there are different stages of fleet planning based on the time frame. The planning sequence consists of the following steps; long-term capacity allocation planning, detailed planning, operational planning and tactical planning (Schmitt and Gollnick 2016). For the long-term capacity allocation planning, the airlines adjust themselves for the amount of aircraft that need to be ordered to either be leased or bought more than a year in advance. This decision depends on the overall market demand for air travel (Schmitt and Gollnick 2016). To find out the market demand airlines retrieve either their traffic data or buy it from other sources. The airlines then use complex data analysis methods to conclude the demand for a particular location that is relevant for them. Based on these market indicators, the airlines order a certain amount of aircraft fleet. The information on the amount of fleet ordered by the airlines is then provided to the aircraft manufacturers to produce and deliver the capacity. The amount of time taken to deliver the ordered fleet is often referred to as aircraft manufacturing delays. The other operational stages done by the airlines until the final day of the flight is not relevant to this thesis therefore are not discussed here in more detail and can be found in (Schmitt and Gollnick 2016).

2.4 Airports

In this section, aspects about airport sub-section of the air transport system are discussed. The discussion involves explaining the functioning and type of airport. In the last part, the effect airports can have on economy is explained.

2.4.1 Functionality and Operations

Airports are the connecting link between air travel passengers, airlines, air navigation service providers and aircraft fuel providers. Airports play an important role in the air transport system by bringing together four of the most important elements of the air transport system. Airports can be defined as the port of transportation from where passengers who wish to travel by air can avail the services provided by the airlines. Airport is the base where the core operation of the air transport system takes place. The airports usually consist of either one or a number of runways that are used for take-off or landing by the airplanes. A number of aircraft can be guided and handled on the airport depending upon the size of the airport. For instance, some airports can have up to 6 parallel runways and can handle up to 1000 aircraft per day (Schmitt and Gollnick 2016). These aircraft are handled at the airport with the help of air navigation service providers (ANSP) which usually operate from a control tower that can watch over the airport. The task of the ANSPs is to coordinate with the pilot and flight staff to schedule take-off, landing and apron slots management. Airports are an essential element in swift functioning of transfer of passenger across border of national regions. This is because airports also consist of the functional and operational section such as passenger boarding areas, baggage treatment functions, security checks and documents and custom check sections. Without strict regulations and protocols at these sections of the airports, the swift and safe functioning of the air transportation would not be possible. The air transport security systems have evolved strongly over the past decades. The rules were revised and made even stricter after the 9/11 incident so as to improve the air transport's image to the general public (Blalock et al. 2007). Apart from the connecting the components of the air transport system, airports also serve as a connecting link between the city and the air travel passengers. Since airports are usually located far off from the main city, therefore, there is a huge requirement for a connecting link between the airports and the city. To solve this problem, airports generally have the infrastructure available for passengers to reach airport by other means of transport such as rail, metro, taxis and busses. It can be argued that this availability of infrastructure at the airports allows the passenger demand to be realized in the first place. Many airports also consist of places for hospitality to help businesses to thrive as it has in built hotels, conference halls and business areas as an integral part. Having conference halls at the airport especially is highly beneficial as organizing generates a high amount of economic impact not only for airlines but also for increased revenue for the airports themselves. Furthermore, airports as a location for major conferences create easy access for the participants. Just like airlines and other aspects air transport system are managed by organizations like IATA and ICAO, airports are managed and organized by the Association of Airports (ACI), located in Montreal, Canada. ACI has more than four regional offices in nearly all continents. Similar to IATA and ATAG, ACI is also a non-profit organization that supports airports worldwide and represents their interest as well as plans for their sustainable future (Schmitt and Gollnick 2016).

In terms of usage fee, the passengers usually have to pay an airport tax fee which is levied on the passengers to use the airport and generally is included in the ticket price. Similarly, airlines that operate at an airport pay airport fee as well. They establish a distinct commercial connection with the airport through negotiations involving time slots for their flights arriving and departing, the range of services provided by the airport, and the associated fees covering landing, take-off, as well as station costs at designated gates or apron positions (Schmitt and Gollnick 2016).

2.4.2 Types of Airports

Airports are mainly classified into three different types namely international, national and military or research airports. The classification of airports is usually done based on size, number of passengers, number of international airlines using it, number of runways and number of employees. A brief explanation of each type is as follows:

2.4.2.1 International Airports

International airports are the type of airports from where a lot of flights embark for international routes. These types of airports are usually large and cater to a large number of passengers travelling across borders. International airports are usually dominated by international airlines as they strive to make these types of airports their hubs. However, this does not mean that domestic airlines do not use these airports as these airports provide feeder flight opportunities for smaller airlines. Naturally, due to the large nature and grandiose operations at the international airports, a large number of employees are required to maintain and run the services. Furthermore, international airports require more security and border checking labour force as the majority of passengers travelling to and from this airport are from across borders. These passengers can either be transit passengers or willing to enter or exit the country but regardless of the intention, border controls and security checks are an extensive part of international airports.

2.4.2.2 National Airport

National airports are the type of airports that are a part of the direct network of point to point network strategy. These types of airports are usually used by LCCs and on certain occasions by FSNCs to carry their passengers to the international hub airports. As compared to international airports, national or domestic airports do not have a large number of passengers. Domestic airports require fewer employees to manage and operate as fewer passengers are traveling. Although the security level requirements are similar to international airports the amount of employees is significantly less as the border control aspects do not exist at the domestic airports.

2.4.2.3 Military Airfields or Research Airports

Military airfields are the type of airfields that are used for military purposes. These areas are highly secured and controlled by military personnel. These areas are used to carry out military training exercises or military airfield bases in the events of war. On the other hand, research airports are the type of airports in which universities, think tanks, research institutes conduct their research. Usually, the research airports have only one airstrip that is used by research organizations to test new technologies and advancements.

2.4.3 Airports and Economic Growth in Regions

It has been found in the literature that airports can be strong economic drivers for its regions because as discussed above airports are a strong driver for job creation either directly or indirectly. Also, airports offer good international connectivity, which is arguably a pre-requisite to attracting tourists, increasing enterprises' attractiveness and generating economic value (Schmitt and Gollnick 2016). Statistics have shown that expansion and growth of existing airports can also create jobs in the area. It has been shown that to support the volume, almost an increase of 1000 jobs is required for every additional 1 million increase in passenger handling at the airports (Schmitt and Gollnick 2016). This is because most of the odd tasks at the airport such as offloading passengers, refuelling the aircraft, water filling, waste deployment and positioning the aircraft at parking positions are required to be done by trained professionals at the airports. Since these are manual tasks and given the current technologies, these tasks are far from being automated in the near future.

Due to the manpower requirements, sometimes the increase in the capacity of the airport can be compared to creating a set of small and medium enterprises. Depending upon the wealth and region of the airport, sometimes the creation of each job at the airport creates one to two additional jobs due to spillover effects (Schmitt and Gollnick 2016). Beyond the employment opportunities generated, the airport assumes a significant role in site selection for new businesses. In the global trade landscape, swift connections are a pivotal consideration for industrial enterprises. Hence, a proficiently operating airport becomes a fundamental requirement for companies contemplating establishment in a particular region. Numerous reports delve into the analysis of specific airports, examining their economic influence on the surrounding areas. The overarching consensus derived from these studies leads to the conclusion that airports play a pivotal and indispensable role in the overall development of both a region and a country (Schmitt and Gollnick 2016).

The complexity of determining the real answer for the development of a specific airport is evident, as highlighted in the (European Commission 2013) study Commissioned by the European Commission in 2011. The study focused on "Airports as drivers of economic success in peripheral regions" (ADES). Some of the key conclusions from the study include the recognition of general accessibility as a crucial locational factor. For remote regions, the presence of airports with sufficient scheduled flights emerges as vital for economic development. Contrary to common assumptions, the primary reason for bottlenecks is often not a lack of infrastructure but rather the absence of scheduled flights to pertinent destinations.

Moreover, the study suggests that the limiting factor for economic prosperity is frequently the availability of qualified manpower. Interestingly, the report advocates for the utilization of larger airports in neighboring regions, reachable within a three-hour window, over establishing a new airport. The report further concludes that not all existing airports are indispensable, proposing the closure of some for efficient land use. The study also underscores the potential for improvements to existing airports to enhance their attractiveness. Ultimately, effective airport policy and strategy are identified as pivotal factors that can significantly influence outcomes.

2.5 Aviation Regulatory Authorities, Environment and Emissions

In this section, the combined discussion of aviation regulatory authorities and environment sub-section are done. These two sub-sections are discussed together because initiatives taken by the regulatory authorities can directly affect the environment and aviation emissions. The section provides an overview of the different types of emissions and the recent growth in emissions due to the growth in the industry. In addition, it contains the different initiatives introduced by the regulatory authorities for a sustainable aviation system in more detail.

2.5.1 Environment: Growth in the Aviation Industry and Emissions

The air transport system has been growing rapidly in the past few years. Its growth has been witnessed by an increase in the number of passengers, traffic, directly employed staff and the value the system adds to the GDP of the world. With the increase in traffic and prosperity, the air transport system has also been responsible for causing negative climate changing effects. This is because the operation of aircraft, support vehicles on the ground and other ground transportation causes emissions. Due to the increase in emissions, the aviation system has recently gotten a lot of attention from the media, climate activists and climate scientists. To understand the harmful effects of the aviation system, its effect has to be discussed in the form of numbers and statistics.

Global aviation traffic has been increasing rapidly in the past four decades and majority of which has been witnessed in Asia and other developing regions as airline companies find market expansion opportunities in these areas. This growth is also expected to continue in the near future because according to Airbus and Boeing 1229 and 1031 more aircraft are ordered for the Asia region (Airbus 2017; Boeing 2018) in the near future. These orders are almost equivalent to 21% and 38% of their total orders, respectively. Furthermore, it is expected by Airbus that almost 41% of the next orders will be from the Asia-Pacific region (Airbus 2017). Due to this growth, there has been a significant increase in carbon dioxide CO_2 emissions. Carbon dioxide is a gas that is released upon the combustion of jet fuel in the jet engine. Historically speaking, the average rate of CO_2 emissions from aviation for the period 1960–2018 is 15 million tons (metric) CO_2 per year. But astonishingly, the average growth rate of CO_2 emissions for the period 2013 to 2018 has been almost 3 times larger than for the period 1960–2018 (Lee et al. 2021) equivalent to 44 million tons (metric) CO_2. The annual growth rate of CO_2 emissions from 1970–2012 has been 2.2% per year but this value has been 5% for the 2013–2018 period (Lee et al. 2021). This has accounted for almost a 27% increase in CO_2 emissions from the aviation system. In the year 2018, worldwide carbon dioxide (CO_2) emissions from aviation surpassed 1,000 million tonnes annually for the first time. The total cumulative emissions from global aviation spanning from 1940 to 2018 amount to 32.6 billion tonnes of CO_2, with roughly 50% of this total emitted in the last two decades. This goes to show that the growth in the industry has increased significantly in the last two decades. This growth has been because of the increase in the entrance of LCCs in the market. Currently, the aviation system accounts for approximately 2.4% of anthropogenic CO_2 emissions across the globe. Within the EU, this share corresponds to 3.6% (EASA 2019). The reason why it's extremely important to discuss these numbers is because carbon dioxide (CO_2) has been identified as one of the major drivers of global warming. CO_2 is considerably harmful to the environment due to its rapidly increasing radiative forcing (RF) effect (Etminan et al. 2016). Furthermore, carbon can stay in the atmosphere for more than 100 years and cause negative environmental effects.

Climate scientists have also warned about the non-CO_2 effects coming from the aviation system. In some cases, the non-CO_2 effects are more harmful than the direct CO_2 effects on the environment. Researchers in the past have debated the aviation CO_2 versus the non-CO_2 effects. It has been the case since the (IPCC 1999) report that the comparison of the radiative forcing of CO_2 with radiative forcing of non-CO_2 has been a major topic. Apart from CO_2 emissions, the combustion of kerosene used as jet fuel produces a large number of other harmful GHGs. Specifically, these are indirect greenhouse gases such as nitrogen oxides (NO_x) or carbon monoxide (CO). These gases have no direct greenhouse effect, but they influence the chemical reaction and formation of other greenhouse gases such as ozone (O_3) and reduction in the abundance of methane gas (CH_4), which are harmful to the earth's atmosphere (NASA 2008; Schmitt and Gollnick 2016). Furthermore, the combustion of kerosene at a higher level of altitude produces water vapor known as contrails. It has been found that contrails promote cloud formation in the sky and as a result of this, the radiation from the sun to the earth is reflected before it leaves the atmosphere thus increasing the overall warming process (Schmitt and Gollnick 2016). Recent studies indicate that emissions at high altitudes can increase the radiative forcing in the environment by a factor of 3 by 2050 (Bock and Burkhardt 2019). Furthermore, aviation aerosols are small soot particles that are usually black and organic carbon. They consist of harmful compounds such as Sulfur (S) and Nitrogen (N). (Lee et al. 2009) in 2005 estimated that the sum of both CO_2 and non-CO_2 effects contributed to about 5% of the overall anthropogenic forcing in the atmosphere of the Earth. Even though the current share of aviation based anthropogenic emissions is quite less at the global level, research has shown that if the industry grows according to the projected industry growth patterns (4–5% annually), their effects can be more drastic in the years to come (Airbus 2019). Moreover, looking at the past growth rates, the number of air passengers carried worldwide in 2016 grew by 6.3% to a record value of almost 4 billion. This growth in the year 2016 was attributed to the 3.1% growth in the global economy and fueled by lower fuel prices. Specifically, the fuel prices were on average 43% lower in 2016 than the previous year (European Commission 2017). Naturally, it is unrealistic to expect this sort of growth in the near future in the industry. But research has suggested that the airline industry can grow fast after its recovery from COVID-19 by 2026 (Gudmundsson et al. 2021; Michelmann et al. 2023). Therefore, from the above numbers, it is clear that the aviation industry is in need of an overhaul change to be able to reduce its environmental impact.

2.5.2 Initiatives for the Sustainable Aviation System of the Future

The air transport industry is at a pivotal point as the decisions taken by the regulators and ATS organizations today will impact the future of the system. As discussed in the section above, the ATS system is in requirement of change to help it move towards a sustainable future. In this section, a set of recent initiatives taken by the regulators, action groups, industry leaders and aviation organizations are discussed in detail.

2.5.2.1 IATA Net Zero Carbon 2050 Resolution

A resolution was passed on 4th October 2021 at the 77th IATA Annual General Meeting in Boston, USA to commit to achieving net zero carbon emission from aviation operations by 2050. According to the IATA, these net zero goals bring the aviation industry in line with the Paris Agreement to limit the global warming scenarios below 2°C. Furthermore, since on the 41st assembly of ICAO in October 2022 had a Long Term Aspirational Goal (LTAG), the governments participating in that meeting now also share the same target goals for aviation's decarbonization (IATA 2022a). Table 2.1 illustrates the milestones for the IATA's strategy in a tabular format.

According to the IATA's calculation based on current projections, it is expected that the air passenger journeys could exceed 10 billion in 2050. Furthermore, in the business as usual case trajectory from 2021–2050, the aviation industry was expected to emit 21.2 gigatons of CO_2 emissions. Therefore, to reach the net zero goal a total of 21.2 gigaton of CO_2 emissions would need to be abated by 2050.

IATA believes that 65% of this reduction will be done with the help of sustainable aviation fuels, 13% from new technologies like electric, hybrid-electric or hydrogen combustion technology, 3% will be achieved from infrastructure and operational efficiencies and 19% of the reduction will be achieved with the help of carbon capture and carbon offsetting strategies. According to IATA, the success of achieving the net zero goal would require highly coordinated and combined efforts of the entire industry and its stakeholders such as airlines, airports, air navigation service providers and manufacturers along with significant government support. To achieve these goals, IATA has formulated four strategic key points. First, the adoption of Sustainable Aviation Fuel (SAF) derived from feedstocks that avoid environmental degradation. Second, investment in cutting-edge aircraft technology as it is one of the key aspects. This would include innovative aerodynamic advancements and alternative propulsion systems such as electric or hydrogen solutions.

Table 2.1 IATA net zero 2050—milestones towards net zero aviation taken from (IATA 2022c)

Date	Amount of CO_2 abatement	Pathway	Action
2025	381 Mt	• 97% Carbon offsets • 2% SAF • 1% Aircraft improvements	• ICAO agree long-term goal for international aviation (2022); • Energy sector commits to at least 6 million tonnes SAF production; • Agreement of full implementation of Article of Paris Agreement
2030	979 Mt	• 93% Carbon offsets • 5% SAF • 2% Aircraft improvements	• Use of 100% SAF in aircraft • Upgrades to deliver fuel efficiency improvements of 0.3% by 2030
2035	1,703 Mt	• 77.5% Carbon offsets • 17.5% SAF • 3% Aircraft improvements • 2% CCUS	• Evolutionary technology achieving 30% reduction in fuel burn • Electric/hydrogen aircraft for regional markets (50–100 seats, 30–90 min flights) become available
2040	3,824 Mt	• 44.5% Carbon offsets • 40% SAF • 7.5% new propulsion technologies • 5% CCUS, • 3% Aircraft improvements	• Feasibility of new aircraft such as blended-wing bodies demonstrated with full-scale working prototypes, electric/hydrogen for short-haul markets (100–150 seats, 45–120 min flights) become available.
2045	6,153 Mt	• 55% SAF, • 24% Carbon offsets • 10% new propulsion technologies • 8% CCUS • 3% Aircraft improvements	• Necessary infrastructure for new energy requirements (low carbon electricity/hydrogen) becomes available

(continued)

Table 2.1 (continued)

Date	Amount of CO_2 abatement	Pathway	Action
2050	8,164 Mt	• 65% SAF • 13% new propulsion technologies • 11% CCUS • 8% Carbon offsets • 3% Aircraft improvements	• Commercially viable annual SAF production of 449 billion liters available

Mt: Mega tonnes ; $\mathbf{CO_2}$: Carbon dioxide ; **CCUS**: Carbon capture utilization and storage

Third, efforts to enhance infrastructure and operational efficiency, with a specific emphasis on advancements in air traffic management. Fourth, approving offsets including the integration of carbon capture and storage technology that would eventually contribute to the overarching strategy aimed at mitigating the environmental impact of aviation activities. To reach the net zero goals IATA has down a concrete plan in the form of milestones. These milestones are the building blocks for the net zero strategy.

2.5.2.2 ACARE Flightpath 2050 Goal for the EU

Considering the harmful effects of aviation on the environment in the future, the Advisory Council for Aeronautics Research in Europe (ACARE) introduced "Flightpath 2050" (European Commission 2017; ACARE 2020). The program was launched in 2012 with the aim to strengthen the European aviation industry while taking ambitious emission reduction targets into account. The goals of the program were divided into three distinct features or sections namely; maintain and extending industrial leadership, protecting the environment and the energy supply and ensuring safety and security. ACARE believed that the European aviation is at the forefront in terms of the dominance of the FSNC and LCC type of carriers across the globe and in a way set the industry standards, it would be a wiser step to strengthen it further.

In light of the goal of this thesis, the most important sub-section of the Flightpath 2050 plan is protecting the environment and the energy supply. Under this sub-section, ACARE laid down a highly ambitious climate reduction goal for the Europe's aviation industry. In other words, the Flightpath 2050 goals mean that, in the year 2050, advancements in technologies and operational procedures have achieved remarkable environmental milestones in the Europe's aviation sector. These innovations would contribute to a 75% reduction in CO_2 emissions per passenger kilometer and furthermore, an ambitious target of a 90% decrease in NO_x emissions. Furthermore, the perceived noise emissions from airborne aircraft should be significantly diminished, registering a notable 65% reduction compared to the capabilities of typical new aircraft in the year 2000 (European Commission 2017; ACARE 2020).

For the sub-section to maintaining and extending industrial leadership, the main goal is that the European aviation industry stands as a formidable global competitor, showcasing unparalleled products and services, and commanding a substantial share exceeding 40% of the international market (European Commission 2017; ACARE 2020). Lastly, for the ensuring safety and security sub-section, the goal is to achieve a safety record of fewer than one accident per ten million commercial aircraft flights by 2050.

2.5.2.3 ATAG Aviation's 2050 Goal

Apart from the major aviation goals set by IATA and ACARE, thc Air Transport Action Group (ATAG) has also come forward with its goals for the future of aviation. The Air Transport Action Group (ATAG) is a not-for-profit association that brings together actors and experts from all parts of the air transport value chain (ATAG 2020a). Similar to IATA, ATAG has also resolved the net zero carbon emissions goal for the aviation industry but its aim is mostly towards increasing efficiency in the existing technology and support for carbon offsetting schemes.

Specifically, according to ATAG, the net zero goal is to be reached according to the 2005 year emission value. Since 2005, aviation emissions reached 650 million tons of CO_2, thus for the net zero goal the emissions would have to be reduced by 325 million tons of CO_2 by 2050 (ATAG 2021). To support the trajectory towards the goal, ATAG and ICAO have made two action plans. First, to achieve an efficiency in the jet engine of 2% per year till 2050. Second, using the carbon offsetting and reduction scheme (CORSIA) to support the industry goal of capping the net emissions at 2020 levels (ATAG 2019). Both of these goals were planned by the ICAO for the international flights and the role of the ATAG would be to support them to achieve these goals.

2.5.3 Global and EU based Policy Measures for Sustainable Aviation of the Future

After discussing the initiatives and goals that have been set for the aviation system, it is important to discuss the different methods through which these goals can be achieved. Scientists, researchers and engineers have found multiple ways through which the ATS can reduce its climate change impact. From a high level view, the ways in which aviation can reduce emissions are; efficiency due to Air Traffic Management (ATM), use of alternative fuel, technological and design improvement and policies for emissions reduction through regulatory framework (IATA 2022a). The goal of this section is to conduct a comprehensive literature review and discuss the policy and regulatory framework initiatives that can or are being taken to reduce the negative climate impact of the aviation system.

2.5.3.1 Carbon Taxes, Fuel Taxes and Penalty Taxes

In the past, many researchers have tried to find methods to discourage or encourage a specific economic activity. Taxes are one method that can help in discouraging a particular economic activity. It has been found in the economic theory that taxes can serve as a market based instrument in environmental policy (Krenek and Schratzenstaller 2016). They are a cost-effective way of attaining environmental goals or discouraging activities that produce emissions. The main rationale behind taxes is that they are instruments to internalize the external cost (Krenek and Schratzenstaller 2016). In terms of aviation, this would mean the cost incurred by the users of air services that are not levied on themselves but are imposed on uninvolved third parties. Just like Pigovian taxes, the aviation taxes aim at the internalization of the externalities which can range from CO_2, NO_X or other kinds of GHG emissions (Pigou 1929). Levying a tax increases the cost of using the air travel marginally and this internalizes the costs. Furthermore, it increases the price which reduces the demand thereby also reducing the emissions.

One of the ways in which taxes can be effective is that they should provide an incentive to reduce fuel use thereby reducing emissions. This could be done by the airlines by using more efficient aircraft, choosing a new fuel mix, or a better technology. This would also mean that the airlines would maximize the aircraft load and minimize fuel use (Krenek and Schratzenstaller 2016). This translates into the reduction of emissions per passenger or reducing the number of flights. This could be particularly true for short-haul flights as they are highly carbon intensive.

Researchers in the past have tried to find the different type of taxes that can be implemented to achieve positive environmental impacts. (Krenek and Schratzenstaller 2016) has described different possible aviation tax designs that can increase environmental effectiveness. Figure 2.3 below shows the diagram of the different types of aviation taxes (Krenek and Schratzenstaller 2016). In the aviation industry, there can be majorly three types of taxes, namely, conventional specific taxes, carbon based taxes, and value added taxes. Conventional taxes can be levied upon seats, flight fuel, or tickets. On the other hand, carbon based taxes are mostly levied on fuel and tickets. Lastly, the value added taxes are laid upon the fuel and tickets. Most of the time, taxes on tickets and flights are levied upon the passengers and other taxes like seats and fuel are to be paid by the airliners. Each of these taxes are discussed in detail as follows.

Conventional taxes

A flight ticket is one of the most common version of conventional tax practice. This type of tax is levied upon the passengers traveling to and from an airport. Many countries apply these taxes in the form of graduated tax rates. In other words, the tax increases with the amount of distance traveled. This is usually divided into three segments, which correspond to short, medium and long-haul flight zones flight ticket tax aims at discouraging the passengers due to the high fee that they have to pay to travel. This sort of tax helps in reducing the demand, especially in the short-haul travel which are the most emissions emitting flights. One of the criticisms of the conventional ticket tax is that the conventional ticket tax does not induce the airlines to reduce their fuel use or to change their fuel mix to a new type of fuel i.e., sustainable aviation fuel (SAF). In other words, the environmental effectiveness of the flight ticket tax is quite limited, in some cases, the environmental effectiveness of the ticket tax can be improved by charging economy flights lower than premium flights as premium flights take up more floor space (Bofinger, Heinrich; Strand 2013).

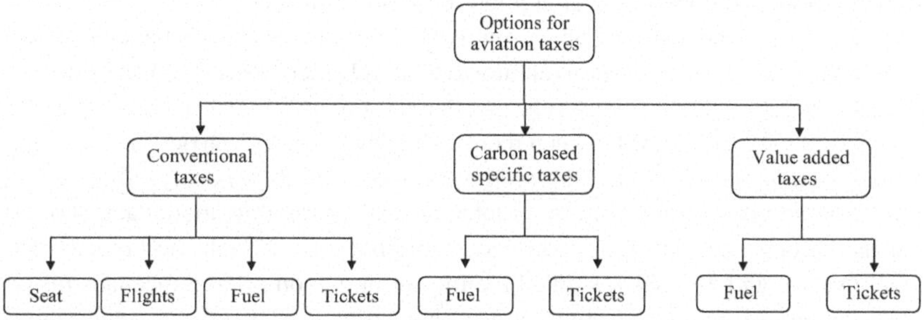

Fig. 2.3 Different types of aviation taxes taken from (Krenek and Schratzenstaller 2016)

A seat tax is a kind of tax, which is levied on the airlines on the amount of seats that are being operated with conventional aircraft. The main motive behind the seat tax is to induce the airlines to maximize their aircraft load as the tax is also applied on the empty seats. (Talwar et al. 2023) used the case of seat tax in their analysis to test for the discouragement of the conventional fleet.

Amongst the other types of conventional taxes, fuel tax is another form of tax that can be imposed in the airline industry. Fuel tax means that the fuel used by the airlines to operate their aircraft will be taxed according to the volume used. This type of taxation has proven itself to be the environmentally most effective one (Krenek and Schratzenstaller 2016). This type of tax incentivizes the airlines to reduce their fuel use by either installing cleaner technology or optimizing flight routes. This would also mean that the airlines would try to maximize their aircraft load and minimize cost per passenger, which may result in a decrease in the supply of flights as well as the demand. This type of tax is not only limited to commercial aviation but can also be extended to freight flights as well. In the past, measures like jet fuel tax have gained attention for their potential to achieve emission reduction goals in the aviation sector. In this area, (González and Hosoda 2016) reviewed the effects of the aviation fuel tax that has been levied on fuel for all domestic flights in Japan since 1972. The authors suggested that the quantity of aviation-related CO_2 emissions can be discouraged by the application of a fuel tax. Similarly, (Fukui and Miyoshi 2017) in their study using the data from the US airline industry examined the effect of an increase in aviation fuel tax on reductions of fuel consumption and carbon emissions. The results suggested that the amount of the reductions would be smaller in the longer term. Initially in their analysis, it was observed that the tax would reduce CO_2 emissions in the US by approximately 0.14–0.18% one year after the tax increase. However, perhaps due to the rebound effect, the reduction would decrease to about 0.008–0.01% after three years. Since all taxation obligations for the use of kerosene as aircraft fuel are currently abolished in the EU, (Faber et al. 2018) explored how a market distortion relating to bilateral agreements could be limited or avoided altogether. Their analysis showed that it is possible for EU member states to tax aviation fuels on flights between them, even when non-EU carriers are enjoying a mutual exemption from fuel tax on those routes. (Tol 2007) analyzed the effect of a kerosene tax with the help of an international tourist flow model and showed that the carbon emission reduction was negligible even though the tax size was large. (Mayor and Tol 2010) studied the impact of carbon pricing mechanisms on emissions in Europe. In this study, it was shown that the UK (air passenger duty) APD reduced more carbon emissions than the tax that was implemented on flights in the Netherlands. (Hofer et al. 2010) applied ticket tax to the US aviation industry and revealed that carbon emissions could decrease proportionally to the magnitude of the tax. (Seetaram et al. 2018) estimated the UK outbound traveler's willingness to pay the tax with distance bands, and found that business class travelers were more willing to pay for long-haul trips.

Carbon based taxes

Apart from the taxes discussed above, carbon based aviation taxes are also another method to environmental impact. The carbon based aviation taxes are different from the conventional ones because they are directly attached to the carbon emissions of a flight. It has been found that this type of taxes is far more effective than the other conventional forms of taxes (Krenek and Schratzenstaller 2016). This is because the carbon-based taxes incentivize the airlines to change their fuel mix or adopt newer technology, it can also be the case that a carbon based fuel tax reduces the demand and the supply of short and long-haul flights. When the carbon based fuel tax is compared to carbon based flight ticket tax, it can be argued that the former is more effective as the latter does not give any incentive for carriers to maximize their aircraft load. One criticism of carbon based tax is that they can be vulnerable to avoidance by the carriers as well as the passengers when implemented unilaterally (Krenek and Schratzenstaller 2016). For example, if the airlines are not able to pass, the fuel tax to passengers, there will not be an impact on the flight demand. Furthermore, if the airlines transfer the carbon base fuel tax to the less price elastic travelers, i.e., business travelers then this type of tax might not be effective. Therefore, for the carbon fuel tax to work, its effect must be translated to the customers who are highly price elastic which are mostly in the low cost carrier segment. In this regard, some of the studies that implemented carbon based taxes to test for their effectiveness are discussed as follows.

In their study, (Dray et al. 2014) used an Aviation Integrated Model (AIM) to implement a carbon tax to accelerate the fleet renewal of airlines. Its impacts were considered holistically, taking into account likely interaction and feedback effects. The author's finding suggested that globally by the year 2050 passenger aviation related CO_2 emissions from a life cycle perspective could be reduced by around one third due to the above policy. Also, (Qiu et al. 2020) investigated an air passenger transport carbon tax incentive policy setting problem, in which the policy includes a carbon tax and competitive tax rebates. Their results indicated that under certain conditions the policy can encourage the airlines to improve fuel consumption. With regard to taxing air travel passengers, (Mayor and Tol 2007) used a model of international flows of tourists to estimate the effect of changes in a boarding tax implemented by the UK government. In their analysis, the authors find that the effects of all proposed variants are small and perverse in increasing carbon dioxide emissions. On the contrary, the authors found that an emission tax had the desired result of reducing emissions. (Falk and Hagsten 2019) investigated the impact of a flight departure tax on air passengers on dynamic panel data for airports in Austria and Germany in the period 2008 to 2016. The authors found that the affected airports observed a reduction of 9% of passengers in the year of the introduction and 5% in the year after. Similarly, in a study conducted by (Krenek and Schratzenstaller 2016), the anticipated revenue from the implementation of a carbon-based flight ticket tax at the EU level and its distribution across EU member states was estimated. Utilizing demand elasticities from (Intervistas 2007), the authors precisely calculated the tax revenues per passenger per route that could have been generated in 2014.

Value added tax (VAT)

Considering from a high level view, the value added tax on flight and fuel can also be regarded as aviation taxes. However, it can be argued that VATs are not environmental taxes whose specific purpose is either to reduce emissions, demand or increase the adoption of new technology by discouraging old ones. The main difference between the above discussed taxes and value added taxes is that VATs are ad valorem tax levied as a fixed percentage on flight tickets or fuel prices rather than a specific predefined amount. One of the advantages of this type of tax is that the value is not degraded due to an increase in inflation. One of the main disadvantages of this type of tax is that it is unclear if they can be specifically used to achieve sustainable outcomes (Krenek and Schratzenstaller 2017).

Challenges and criticism for implementing taxes

Although aviation taxes tend to work in theory, there is a different story in terms of practice. It has been found that it is difficult to implement taxes in the aviation industry. One of the problems to implement taxes in aviation is due to the existence of cross border externalities. In other words, aviation is a highly international industry. This means that the taxes by different countries cannot be the same as usually countries do not agree on the same terms. Furthermore, many times the taxes are set, inefficiently low, and are set without coordination between countries. Another problem while implementing the aviation taxes is that there can be a downward tax competition. Countries in general, do not have a clear incentive to tax the aviation industry even if a common agreement is achieved. The countries might have an incentive to slowly reduce their taxes over time. Due to this, some airline companies can also indulge in tax avoidance measures, for example, passengers traveling to airports in non-taxing countries or airlines refueling abroad. This type of practice in turn could also reduce the positive impact of taxes on the environment.

2.5.3.2 Quotas, Fleet Restriction and Fuel share Obligations

Apart from using taxes as a discouraging mechanism, the regulators of the system can also use other methods to encourage the use of environmentally friendly technologies in the air transport system. One of the methods to do this is by using quotas, fuel share obligations and fleet restrictions. Quotas, fleet restrictions and fuel share obligations simply mean that a certain percentage of fuel or fleet must be operated by technology that reduces the environmental impact of air travel. These could be new technologies like electric or hybrid electric aircraft, hydrogen combustion or hydrogen electric aircraft or sustainable aviation fuels SAFs. There have been examples of such types of regulations seen in the past to drive the ATS towards a sustainable future. (Gössling and Lyle 2021) conducted a study about identifying 30 feed-in quota type policies to decarbonize aviation. The authors found that a single feed-in quota forcing the aviation sector to use 100% non-biogenic synthetic fuels could theoretically fully decarbonize aviation by 2050. In the real world, an example of one such approach can be seen in Norway. The Norwegian authorities have recently imposed a requirement that 0.5% of aviation fuel in the country must be sustainable by the year 2020, with a subsequent increase to 30% by 2030 (World Economic Forum 2020). Furthermore, the authorities have also stated their desire to make all short-haul flights within the country to be 100% electric by 2040 (Avinor 2018; Mckinsey 2020). Not only Norway, but Sweden as well recently has committed to being a leading nation in electric aviation development by making it easier for companies to pursue innovation. Sweden also plans to make their domestic flights completely fossil free and has committed to achieving the usage of 100% biofuels for domestic aviation by 2030 (Fossilfritt Sverige 2021; Dagens industri 2022). The country has planned a similar vision for all international flights leaving Sweden by 2045, and for this, they plan to use both hydrogen and electric flights (Larsson et al. 2019). Also, in their efforts to promote electric aircraft, Finland has unveiled their plan for electric aircraft routes across the Lapland region (yle 2022).

The European Union has also taken some steps in this direction to uptake or increase the usage of sustainable aviation fuels (SAFs) in the future in the EU. This has been done with the help of the "Fit for 55 plan". Under the "Fit for 55 package", a clause named "Refuel EU aviation" is introduced for the European aviation system (European commission 2021a). Under the refuel EU aviation, there is an obligation on the fuel suppliers to blend increasing levels of SAF and ensure availability at EU airports for airlines to uplift the usage of SAF. In terms of usage by the airlines, the obligation for SAF share requirements starts at 2% from the year 2025 and increasing every five years. This means that the share should be 5% of SAF by 2030, 20% by 2035, 38% by 2040 and 63%

percent by 2050. In this plan, there are some incentives introduced to force the airlines to make use of SAF in the future. In other words, to ensure that the SAF blending and uplifting obligations are met, the EU commissions define criteria for noncompliance penalties. The member state authorities will propose penalties for fuel suppliers and airlines if the specified requirements are not met.

This sort of encouragement and discouragement of regulations can force the system to move towards new cleaner, technology that is good for the environment. (Talwar et al. 2023) used the combination of policies to test the adoption of electric aircraft in the EU. The authors tested the implications of introducing a fleet restriction policy, along with a jet fuel tax and seat tax policy to observe the effects of an increase in the adoption of electric aircraft. Such sort of policy regulations can be another method to help the air transport system move towards sustainability.

2.6 Aircraft Fuel Producers

Aircraft fuel producers provide fuel to the airline companies at the airport to operate their aircraft. In the past few years, there have been different types of fuels that have been discovered to promote the sustainability of the air transport system. One of these types of fuels is sustainable aviation fuel. Sustainable aviation fuels are biofuels of different types and they are made from different feedstock. Furthermore, depending upon the type of method used for their production, they have different emission reduction potential from a life cycle emissions perspective as compared to the normal jet fuel. Since other type of fuels, such as hydrogen are not yet fully developed or produced because of the lack of development in the hydrogen combustion technology, they are not discussed in the thesis. In this section, the aspects related to the different types of sustainable aviation fuels (SAFs) such as their emission reduction potential, and feedstock availability to be produced at a mass scale are explained in detail.

Table 2.2 Type of SAFs based on their production route and GHG emissions reduction potential

SAF production route	Feedstock	GHG emission reduction potential compared to jet fuel in%*
HEFA	• Used cooking oil • vegetable oil • Animal fats • Algae	38–70%
Fischer-Tropsch synthesis (FT—Syn)	• Municipal solid waste (MSW) • Forestry residue • Agriculture residue • Energy crops	85–93%
Alcohol to Jet (ATJ)	• Methanol • Ethanol • Butanol	38–71%
Power to liquid (PtL)	• Hydrogen • Electricity • Carbon captured from air or Bioenergy Carbon Capture and Storage (BECCS)	Up to 99%

* : Based on life cycle assessment; Source: (Tanzil et al. 2021; Ng et al. 2021; Shahriar and Khanal 2022)

2.6.1 Sustainable Aviation Fuels

The term sustainable aviation fuels (SAFs) has not been clearly defined but is commonly used to describe an extensive range of drop-in kerosene alternatives that can be derived from a multitude of feedstocks (Cabrera and Melo de Sousa 2022). SAFs can be used in place of conventional jet fuel and do not require any structural changes to the aircraft or to airport fueling infrastructure. Various bioconversion technologies exist for the production of SAFs, with some being well-established methods, while others are relatively new and require further advancements. Irrespective of the manufacturing process, SAF (when blended with traditional jet fuel) must meet the stringent requirements set forth by certifying organizations i.e., American Society for Testing and Materials (ASTM) to be considered as a drop-in fuel (Shahriar and Khanal 2022). The major SAF production pathways include Fischer Tropsch synthesis (FT-SPK), hydroprocessed fatty acid esters and fatty acids (HEFA), alcohol to jet (ATJ) and power to liquid (PtL) (Roland Berger 2020a; Zhang et al. 2020; Tanzil et al. 2021; IATA

2022b; Cabrera and Melo de Sousa 2022; Hsu et al. 2022). The major types of feedstocks suitable for SAF production can be grouped into five categories oil, sugar, alcohol, starch, and lignocellulosic feedstocks (Cabrera and Melo de Sousa 2022). Although it is found that the combustion of SAFs produces almost similar amounts of CO_2 emissions when combusted in a jet engine, major emissions reductions can be achieved by reducing emissions from the production side (Barke et al. 2022). Studies show that SAFs can decrease greenhouse gas emissions (GHG) by exceeding 60% when compared to conventional jet fuel (De Jong et al. 2017). Furthermore, it is found that the combustion of SAFs lowers other emissions such as soot, sulfur oxides, aromatic compounds and ultrafine particles resulting in less particulate matter and NO_x released during combustion. This reduction also helps in mitigating the climate impact of contrails (Barke et al. 2022; Cabrera and Melo de Sousa 2022). In some cases, the combustion of SAF offers a decrease in fuel consumption and CO emissions (Gawron et al. 2020). The different types of SAF production pathways that are certified by ASTM and their emission reduction potential and feedstock are shown in Table 2.2.

2.6.2 Hydroprocessed Fatty Acid Esters and Fatty Acids (HEFA)

The HEFA process to produce SAF is an accepted bioconversion process under specification ASTM D7566 incorporated in June 2011. SAF produced through HEFA route has a maximum allowable blend ratio of 50% with the jet fuel. In the HEFA route, the oils are transformed from lipids to hydrocarbons in a four-stage process (De Jong et al. 2017; Bauen et al. 2020; Ng et al. 2021; Cabrera and Melo de Sousa 2022). Producing SAF from the HEFA pathway can be done with feedstock such as vegetable oil, animal fats, used cooking oil, algae, and raw materials from biomass (Hsu et al. 2022; Shahriar and Khanal 2022). Research shows that HEFA technology is currently the most economically competitive and technically the most mature pathway available.

This is mainly due to its high fuel yield which is about 86–91% of feedstock converted to usable fuel (Tanzil et al. 2021; Ng et al. 2021; Shahriar and Khanal 2022). In terms of emissions reduction potential, the HEFA process gives about 16.5–47 g CO_2 eq/MJ compared to the fossil fuel baseline of 89–94 g CO_2eq/MJ. This means that the HEFA process provides about 38–70% life-cycle based GHG emissions reduction compared to conventional jet fuel (De Jong et al. 2017; Hsu et al. 2022; Shahriar and Khanal 2022).

2.6.3 Fischer-tropsch Synthesis

The Fischer–Tropsch process to produce SAF was the first to be approved by ASTM in 2009. The Fischer-Tropsch (FT) synthesis, a method for generating liquid hydrocarbons from coal, was pioneered by Franz Fischer and Hans Tropsch in the 1920s. This process has been firmly established and is widely recognized as a reliable and effective method for producing hydrocarbons (De Jong et al. 2017; Bauen et al. 2020; Ng et al. 2021; Cabrera and Melo de Sousa 2022; Shahriar and Khanal 2022). At present, the American Society for Testing and Materials (ASTM) has certified FT-SPK (Synthetic Paraffinic Kerosine) and FT-SKA (Synthetic Paraffinic Kerosine with Aromatics) as per specifications D7566-Annex 1 and D7566-Annex 4, respectively. These certifications were incorporated in September 2009 and November 2015, respectively. It has been established that the maximum permissible blend ratio for SAF produced through the FT synthesis process is 50% (IATA 2022b). SAF from the FT synthesis route can be produced from lignocellulose biomass. Lignocellulose biomass comprises a wide range of feedstock to produce jet fuel through a thermochemical or biological route. The Fischer-Tropsch (FT) process can be integrated with a variety of biomass conversion techniques, including gasification, pyrolysis, and liquefaction, for the generation of synthetic fuel (Ng et al. 2021). Presently, the Fischer-Tropsch (FT) process is considered to be the most environmentally friendly method for producing SAFs. This translates to greenhouse gas (GHG) emissions, with a range of 3.6 to 22 g CO_2 eq/MJ, depending on the type of feedstock utilized which is equivalent to 85% to 93% emissions reduction compared to conventional jet fuel (De Jong et al. 2017). This reduced GHG emissions impact can be attributed to the utilization of lignocellulosic biomass as a feedstock in the FT process (Shahriar and Khanal 2022).

2.6.4 Alcohol to Jet (ATJ)

The alcohol to jet process (ATJ) consists of the conversion of shorter chain alcohol such as methanol, ethanol and butanol into longer chain hydrocarbon (De Jong et al. 2017; Ng et al. 2021). In this regard, mainly two alcohol types (isobutanol and ethanol) are certified for SAF production according to the ASTM specifications (Shahriar and Khanal 2022). Isobutanol to jet fuel is qualified under ASTM D7566 annex 5 and was added in April 2016. Ethanol was included as an ATJ feedstock was included in April 2018. The maximum blending limit for the ATJ process is 50% (IATA 2022b). The ATJ process has a broad range of feedstock options to produce SAF as it can use sources such as sugar or alcohol solutions derived from sugary and starchy biomass and lignocellulosic biomass sources such as corn grain, switchgrass, and sugarcane (Bauen et al. 2020; Shahriar and Khanal 2022). The ATJ offers 38–71% GHG emissions reduction potential compared to conventional jet fuel (De Jong et al. 2017). The substantial variation in greenhouse gas (GHG) emissions is due to the performance of the ATJ pathway technology and methodology. These factors impact the energy requirements and feedstock-to-fuel conversion efficiency, thereby influencing the GHG emissions level associated with a specific feedstock (Shahriar and Khanal 2022).

2.6.5 Power-to-liquid (PtL)

The Power-to-Liquid (PtL) route involves the use of electrolysis to produce hydrogen, here hydrogen ideally should be produced through green means, and the combination of carbon through carbon capture to synthesize hydrocarbons. There are two established pathways for the production of PtL fuels: the Fischer-Tropsch (FT) and methanol pathways. The total yield of hydrocarbons produced by both Power-to-Liquid (PtL) routes is similar, with an estimated range of 17% to 18%. The yield of SAF produced through the Fischer-Tropsch process, when optimized for aviation fuel, is 60%, which may be increased to up to 70% with further advancements in the technological process (Bauen et al. 2020; Cabrera and Melo de Sousa 2022). The American Society for Testing and Materials (ASTM) imposes no restrictions on the origin of feedstock for Fischer-Tropsch (FT) fuels that can be used as drop-in fuels, and thus, the utilization of this pathway in commercial aircraft through blends up to 50% is already possible. The carbon used in the Power-to-Liquid (PtL) process for the production of synthetic fuel can be obtained from three sources: (1) industrial waste gases produced from the burning of fossil fuels such as coal or gas, (2) sustainable biomass through Bioenergy Carbon Capture and Storage (BECCS), and (3) Direct Air Capture (DAC), a process that captures carbon dioxide directly from the atmosphere. The use of renewable electricity throughout the production process of power-to-liquid fuels results in greenhouse gas (GHG) emission reductions of up to 99% compared to conventional fossil jet fuel, regardless of the source of carbon dioxide used. Due to the theoretically unlimited production capacity of PtL methods, this pathway is expected to play a significant role in decarbonization efforts in aviation once technological barriers are overcome (World Economic Forum 2020).

2.6.6 Feedstock Limitations

Despise the increasing popularity of SAFs, its production in 2019 constituted approximately 1% of the total demand for jet fuel. From the years 2013 to 2015, the average global production of SAF was 0.29 million liters per annum. But from 2016 to 2018 the production experienced an increase, reaching 6.45 million liters per annum. In the near future, it is projected that the annual global production of SAF can serve 2% of fuel demand by 2025 and between 2.5%–6.5% by 2030 (ATAG 2020b; ABBB 2021; World Economic forum 2021). After 2030, it is expected that SAF production can reach 6.5 million metric tons equivalent to 8 billion liters by 2032 (Shahriar and Khanal 2022). There exist several obstacles hindering the growth and expansion of the market for SAFs after 2030 but the major problems are the lack of maturity of technologies of the pathways, unavailability of feedstock quantities and not enough support from local governments and international organizations to accelerate and facilitate the fossil to renewable fuels. The confluence of these challenges has a detrimental impact on the cost competitiveness of SAFs, and if these concerns remain unmitigated, it could impede the progress of decarbonization efforts and hinder the industry from achieving its established objectives for the forthcoming decades (Cabrera and Melo de Sousa 2022). Researchers in the past have tried to find the upper limit of availability of feedstocks but there seems to be mixed evidence. Research showed that the production of biofuels primarily relies on simple sugars obtained from agricultural starches in the United States and sugarcane in Brazil as feedstocks. The Association of Southeast Asian Nations (ASEAN), consisting of ten member states, possesses the ability to supply approximately 500 million tons of biomass annually, with palm oil being the most significant contributor originating from Indonesia and Malaysia. Furthermore, Asian nations export used cooking oil, a highly valuable low-carbon feedstock, to the US and Europe for the production of SAFs. A recent study estimated that the major Asian exporters, including China, India, Indonesia, Japan, Malaysia, and the Republic of Korea, possess the capability to provide 9000 kilotons of used cooking oil annually (Shahriar and Khanal 2022). In another study, it was found that sustainable oils and lipids which are feedstocks suitable for HEFA processes can be available to produce 85 million metric tons of SAF. These include waste and residue lipids, oil trees on degraded land and oil cover crops. The same study also revealed that cellulosic biomass, a type of feedstock that is suitable for FT synthesis, has the potential to yield up to 405 million metric tons of SAF. This includes sources such as cellulosic cover crops, agricultural residues, forest residues, waste generated from wood processing, and municipal solid waste. Overall, it has been found that 500 million metric tons of SAF annually can be available from waste Feedstock by 2030 (World Economic Forum 2020). In another report, it was found that by 2050 200 Million tons per year of SAF could be produced outside of PtL production which is theoretically unlimited subject to the price of SAF through that pathway. The above discussion implies that to produce such large amounts of SAF a lot of refineries would be required. In one study, it was found that reducing 15% of emissions by 2050 would require the construction of approximately 60 new bio-refineries annually (Staples et al. 2018). In another study, it is found that up to 5,000–7,000 facilities may be required to produce sufficient SAF to achieve the climate targets of the aviation industry. In 2050, the capacity of an average facility is estimated to be roughly 120 million liters (100,000 tons), producing 83 million liters (65,000 tons) of SAF per annum (ATAG 2020b).

2.7 Aircraft Manufacturers

Aircraft manufacturers are one of the most important components of the air transport system. Aircraft manufacturers develop, maintain and manufacture aircraft that are ultimately used by airlines for their fleet operations. The aircraft manufacturer can potentially be the main instigator for change in the system as they can develop new emissions-free technologies that can be used by the airlines and ultimately reduce emissions. In this section, the different technologies that are being developed in the field of air transport system are discussed in detail. Furthermore, in this section, the challenges, and the difficulties for making the new technologies a reality are also discussed. In other words, this section provides an extensive overview of future possibilities from the aircraft manufacturers sub-section to help reduce the emissions from aviation system.

2.7.1 Electric Aircraft Technology

One of the technologies that have shown potential for the future air transport system is the electric aircraft propulsion technology. Recently, this technology has been researched extensively for its implication in the air transport system. This trend can be due to the increased electrification of vehicles in the automobile sector. The origins of electric aircraft trace back to 1972, marked by the flight of a manned aircraft powered by a nickel-cadmium (Ni-Cd) battery. However, due to the low energy density of the battery and the low performing electric electrical components, the concept could not become a reality (Sahoo et al. 2020). In the 1980s, as electrical component technology advanced, the interest in developing battery-powered electric aircraft began to emerge. One of the motivations behind the electrification of the propulsion system is to make the propulsion more efficient compared to conventional fuel propulsion systems. Scientists believe that this can happen with the help of efficient energy conversion and transmission systems (Sahoo et al. 2020). Another motivation to work towards the electrification of aircraft is to achieve fully electric flight at adequate aircraft weight characteristics to maximize the emission reduction in the use phase (Karpuk and Elham 2021). There have been innumerable different philosophies and studies about the electrification of aviation, but amongst them the most important, four primary domains are: fully electric aircraft, turbo electric aircraft, hybrid electric aircraft and distributed electric propulsion system design. The extent to which the electrical energy source is used in the design is based on the electrical powertrain arrangement in the configuration. Keeping in mind the scope of this thesis only fully electric aircraft is discussed in more detail.

Table 2.3 Parameter difference between CE-Liner and Boeing 737-3+ taken from (Hornung et al. 2013)

Aircraft properties	Units	Ce-Liner	B787-3+	Δ (B787-3+)
MTOW	Kg	109300	73700	+49%
PAX	Kg/PAX	314	253	+24%
Max Energy(Fuel)Weight/MTOW	%	27.5	24.3	+13.2%
Distance range with max PAX	Nm	900 nm		

Δ: change with respect to ; nm: nautical mile ; PAX : Passengers

2.7.1.1 Fully Electric Aircraft Technology

A fully electric aircraft is a type of aircraft system that relies upon a battery or some other means of electric energy source as a sole means of power to the propulsion system. One of the features of these types of designs is that they have a highly efficient conversion system and it is the only type of configuration that has the potential for zero in-flight emissions (Sahoo et al. 2020). Furthermore, this type of aircraft could also provide lower in-flight noise emissions during operations. Due to its emissions, noise and efficiency advantages, there have been authors and aerospace research organizations in the past that have assessed the possibility of a commercially viable short-range electric aircraft.

In 2012, Bauhaus Luftfahrt developed a conceptual electrically operated aircraft named "Ce-Liner". The aircraft is completely electrically operated with almost 190 seats. It is expected that the aircraft can have a range of 900 nm (1666 km). It was found that in order to operate this type of aircraft, batteries with a specific energy of 1.7 kWh per kilogram and 1.2 kWh per kilogram would be required. Furthermore, it was also found that for safety measures the correct amount of battery technology would be targeted to 2 kWh per kilogram (Hornung et al. 2013; Isikveren and Schmidt 2014; Schmidt et al. 2014). This type of aircraft consisted of high temperature superconducting electric motors (HTS).

It was assumed that this aircraft would be only possible for technologies after the year 2035 where battery technologies would be able to reach the 2 kWh per kilogram mark. Table 2.3 shows the overview of the difference in parameters of CE-Liner as compared to a Boeing 787-3 aircraft taken from (Hornung et al. 2013). In terms of weight, this concept had an MTOW of 109.3 tons. This was found to be significantly higher than the accepted conventional aircraft design with 2035 technological levels. Specifically, it was 49% higher than a conventional aircraft.

One of the greatest advantages of this type of aircraft was the electric motor efficiency as compared to the piston engines or gas turbine engines. Researchers found that after comparing all efficiencies of conventional gas turbines from fuel to thrust, the electrical propulsion system was 65% more efficient than the conventional aircraft. Furthermore, the HTS electrical and propulsion system used for this type of aircraft provided almost 98% to 99% efficiencies. Although there were a great number of advantages for this type of aircraft, many researchers criticized the aggressive amount of assumptions for the battery technology for the CE-liner. Apart from the research done on CE-liner, (Schäfer et al. 2019) conducted a technological, economic and environmental assessment of an all-electric aircraft. The authors mentioned in their study that depending on the specific design and quantity of batteries carried, a battery pack with a specific energy of 800 Wh/kg would be required for Airbus A320/Boeing 737 sized aircraft to be capable of up to 600 nm (1,111 km) missions. In the study, it was stated that a smaller first-generation all-electric aircraft with a battery-pack specific energy of 800 Wh/kg and a range of 400 NM (741 km) would be economically feasible only with battery costs of around US $100/kWh or less. Also, for it to be operationally attractive for the companies, policies that provide a high amount of subvention of electricity would be required. (Gnadt et al. 2019) developed a fully electric 180 passenger aircraft based on a conventionally powered Airbus A320neo aircraft design. Their analysis suggested that an all-electric aircraft AEA-800 comparable to the A320neo configuration with 180 passenger capacity and a battery pack with a specific energy of 800 Wh/kg could provide a 500 nm (926km) range. In terms of efficiencies, the transmission of electricity and distributional efficiency was approximately 93% for this type of aircraft. Furthermore, the charging efficiency was assumed to be 95% which was typical for lithium ion batteries (Gnadt 2018). (Karpuk and Elham 2021) developed a regional fully electric aircraft comparable to the conventional ATR-72 design named SE^{2}A SR aircraft. Their results showed that an all-electric aircraft with new airframe technologies might be designed with a maximum take-off weight increase of 50% requiring a battery pack with a specific energy of 700 Wh/kg. Table 2.4 shows the detailed specifications of this aircraft. This particular design was modeled to serve a 500 nm (926km) range. However, it was found that depending on the battery energy density this aircraft had an increase in direct operating cost (DOC) ranging from 43% to 30% which makes these aircraft less attractive for the industry.

Table 2.5 below shows the different costs of the different designs of the SE^{2}A SR regional aircraft as compared to the ATR-72 design. In the table, there are mainly three types of designs that are developed by the authors. Specifically, these are DOC optimal design, DOC optimal design with no folding and the battery optimal design. These three designs of the same aircraft are then compared to the ATR-72 design for costs per flight in euros. It can be inferred from the table that the DOC for the optimized aircraft was reduced by €124 per flight.

Table 2.4 Top level requirement for SE^{2}A SR aircraft taken from (Karpuk and Elham 2021)

Requirement	Units	Value
Design range with max PAX	Km	926
Maximum payload	Kg	7500
Cruise Mach number		0.42
Service ceiling	m	7620
Landing distance	m	1100

Table 2.5 Comparison of parameters between different types of SE^2A SR aircraft designs with ATR-72 taken from (Karpuk and Elham 2021)

Parameter	Units	Initial design	DOC optimal design	DOC optimal design (no folding)	Battery optimal design	ATR—72
MTOM	Kg	35,369	34,441	33,649	34,783	23,000
Battery weight	Kg	12,783	12,262	11,527	10,379	–
DOC	EUR/Flight	9603	9497	9397	9838	6740
mCO_2	Kg	2045	1961	1844	1660	11,838

Furthermore, the difference between the CO_2 emissions for the initial and the DOC-optimized design is equal to 84 kg per flight. According to the author, this amount of difference is not substantial (Karpuk and Elham 2021). In the design with no folding, it was found that the direct operating cost (DOC) was reduced by 206 euros per flight compared to the initial configuration. This indicated that there was a strong effect of folding on the aircraft (Karpuk and Elham 2021). Furthermore, the emission level was reduced by 201 kg as compared to the initial design.

Comparing the battery minimization strategy with other forms of design, it was found that the direct operating cost increased by €235 per flight as compared to the initial design or €359 per flight as compared to the optimized configurations. The battery weight was reduced by 2404 kg compared to the initial design and 1883 kg compared to the DOC-optimized design of the aircraft.

In terms of emission levels, it was found that the emissions reduced by 385 kg compared to the initial configurations. If these designs are compared to the ATR-72 reference aircraft it was found that the most cost-effective aircraft version had a 41% increase in DOC, furthermore the DOC for battery optimal configuration was increased by 46%. Also, the DOC efficient version had the most reduction in CO_2 emissions equivalent to 83% and the battery optimal configuration had 86% emission reduction as compared to ATR-72.

Table 2.6 Battery specific energy requirement according to aircraft type

Aircraft type	Propulsion	Passengers	Battery specific energy requirement
General aviation	Fully electric	2–3	250 Wh/kg
General aviation	Fully electric	4–10	400 Wh/kg
Regional aircraft (ATR-72 comparable)	Fully electric	50–70	600–700 Wh/kg
Short-haul aircraft	Hybrid electric	100–150	600–700 Wh/kg
Short-haul aircraft	Fully electric	180 +	750 + Wh/kg

Source: (Sahoo et al. 2020; Karpuk and Elham 2021)

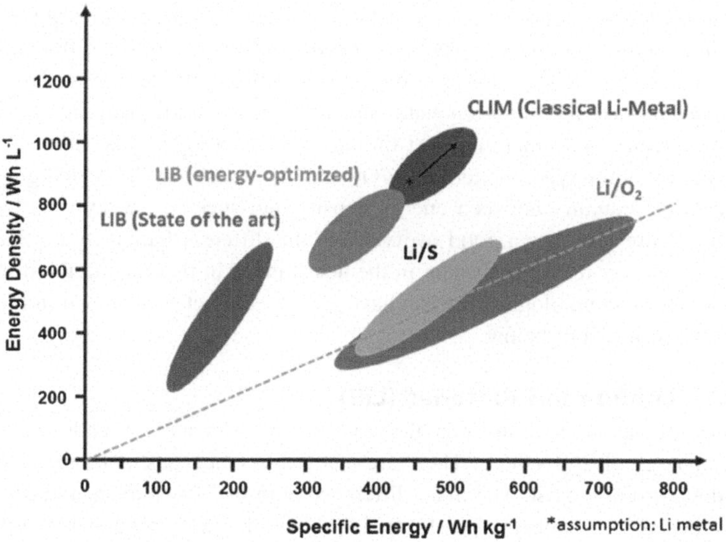

Fig. 2.4 Comparison between energy density and specific energy of the different batteries taken from (Placke et al. 2017)

From the above discussions, it can be inferred that with the introduction of batteries, there is a large increase in the weight of the aircraft. Furthermore, the electric aircraft seem to be more expensive to operate as compared to the conventional aircraft. Lastly, even though research suggests that electric aircraft can largely reduce in-flight emissions, for them to become a reality, a battery with a high specific energy density of 800 Wh/kg and beyond would be required.

2.7.2 Battery Requirements and Challenges

It can be inferred from the above discussion that battery technology is the key to the development of fully electric aircraft. The higher the specific energy of batteries, the larger the planes can be and the further they can fly (Posada 2017). In terms of the minimum specific energy of battery requirements to operate, the aircraft researchers have found that the different aircraft classes would require different types of batteries. The specific requirements of battery specific energy for different types of aircraft are shown are Table 2.6. It has been found that the historical long-term rate of increase in battery specific energy has been around 3% per year. This increase in general has facilitated the doubling of the specific energy every 23 years. Although it has been found that since the year 2000, battery specific energy has increased at the rate of 4% per year, therefore, the continuous increase could be hard to achieve. Figure 2.4 taken over from (Placke et al. 2017) shows the energy density vs specific energy plot of the different types of batteries. In the figure, comparison of current lithium-ion batteries (dark blue) to energy-optimized lithium-ion batteries (light blue), traditional lithium-metal batteries (green), and emerging post-lithium-ion technologies like lithium/sulfur (Li/S) and lithium/oxygen batteries (Li/O_2) can be observed. The dotted gray line represents the equality between energy density and specific energy (Wh/L=Wh/kg). This figure provides a good overview of the different battery types that can be used for the air transport system in the near future. In this regard, the different types of battery technologies that can have high specific energy shown in Fig. 2.4 are discussed in detail below.

2.7.2.1 Lithium Ion Batteries (LIB)

Currently, it has been found that the lithium ion cells have already reached a specific energy of 260 Wh/kg. There are many different types of lithium ion batteries that currently exist. The main difference is in the way anodes and cathodes are composed in a battery type. Table 2.7 shows the different types of lithium, ion batteries with their chemical description and usage. In the table, it can be observed that the early lithium ion batteries used LCO, which was used in electric vehicles. But nowadays the batteries are mostly using NCA, NMC, and LFP type of battery chemistries. One of the highest forms of specific energy battery available commercially is made by Panasonic. The Tesla battery pack consists of these Panasonic NCA models. It has been found that the batteries with LFP type of chemistries are relatively low on costs and offer 26 charging cycles higher than the others. The only problem with the LFP type of batteries is that they are low on specific energy. In terms of lithium ion batteries, it is believed that

NMC type batteries will dominate over the next decade (Gnadt et al. 2019). Amprius Technologies, a battery manufacturer, has introduced its latest lithium-ion cells boasting a high-energy density of 450-Wh/kg and 1150-Wh/L. These cells, utilizing the company's Si-Nanowire anode technology, mark a significant advancement compared to the widely available 300-Wh/kg batteries. Notably, this innovation builds upon the 405-Wh/kg product that was unveiled in 2022 (Future Flight 2022). With the development of the batteries, it is expected that by 2025 a battery capacity of 400 to 500 Wh/kg cells could be achieved.

Table 2.7 Different type of lithium ion batteries their chemical description and usage taken from (Gnadt 2018)

Li-ion type	Chemical description	Specific capacity (Ah/kg)		Electric vehicle application
		Typical	Theoretical	
LCO	Lithium cobalt oxide $LiCoO_2$	150	274	N/A Original B787 battery
LMO	Lithium manganese $LiMn_2O_4$	120	148	2011/2013 Nissan Leaf 2012 Mitsubishi i-MiEV 2013 Ford Focus EV
NCA	Lithium nickel cobalt aluminum oxide $LiNi_{0.8}Co_{0.15}Al_{0.05}O_2$	190	278	2014 Tesla Model S 2015 Mercedes Benz B class ED 2016 Tesla Model X
NMC	Lithium manganese cobalt oxide $LiNi_xMn_yCo_{1-x-y}O_2$	170	278 (NMC 333)	2014 BMW i3 2015 Chevrolet Spark EV 2017 Nissan Leaf, Tesla Model 3
LFP	Lithium iron phosphate $LiFePO_4$	160	170	2011 Chevrolet Spark EV eBuses

Challenges

The current type of technology that exists today has the maximum theoretical weight effective energy limited to 600 to 700 Wh/kg, and in specific cases of higher value, such as 1000 Wh/kg (Staack et al. 2021). However, researchers argue that the increase that has been witnessed in the past cannot continue in the future. This is because for it to improve further a new type of cell chemistry technology would have to be innovated to increase the specific energy in the next 20 years (Schäfer et al. 2019). Recently, there have been studies that assessed the costs and development of potential future battery systems that can suit the needs of future aviation. The discussions in the literature include lithium-sulpur (LSB) and lithium-air batteries (LAB) as potential future batteries because of their higher pack level/theoretical specific energy (Placke et al. 2017; Duffner et al. 2021; Mauler et al. 2021).

2.7.2.2 Lithium Sulpur Battery (LSB) Technology

Lithium sulfur batteries or Li-S batteries feature a sulfur cathode along with the lithium anode (Gnadt 2018). These types of batteries use nano-structured sulfur with high amounts of conducive carbon as the positive active material while lithium metal is used as the negative active material (Placke et al. 2017). Li-S batteries require a high amount of electrolytes which reduces the attainable energy density. But one advantage of this type of battery is that Sulpur is generally abundant and nontoxic so the battery chemistry could theoretically offer high economies of scale and environmental friendliness. It has been found that these types of batteries have a specific energy of 2567 Wh/kg. It is envisioned that LSB battery technology with its improvement in electrolytes, cathode and anode materials would be commercially able to achieve a cells level specific energy of approximately 500 to 650 Wh/kg in the next 15 years. Furthermore, it could reach 800 to 950 Wh/kg specific density in the next 30 years. Currently, the highest development of this technology has demonstrated 500 Wh/kg specific energy and with 1500 cycles in a laboratory experiment (Sahoo et al. 2020).

Challenges

It has been found that LSBs provide low-cost advantages and high theoretical specific energy (currently > 400 Wh/kg), although they cannot be commercialized successfully at a large scale as of now. One of the main issues for LSB is that they have a poor rate of performance and a relatively low cycle life. It has been generally agreed upon in the literature that LSBs will have benefits compared to lithium-ion batteries (LIBs) for applications where high specific energy is necessary (e.g., aviation) and limited cycle life can be tolerated (e.g., military applications) (Placke et al. 2017).

2.7.2.3 Lithium Air Batteries (LAB) Technology

Lithium air batteries are fundamentally different from all the previous types of batteries discussed above. The LAB batteries involve the use of oxygen as a gaseous positive material which ideally can be taken directly from the atmosphere. The LAB uses an oxygen air permeable carbon cathode with a porosity of around 80% (Sahoo et al. 2020). In these types of batteries, lithium is used as the negative active material. The batteries consist of both aqueous and non-aqueous electrolyte formations, but it has been found that the latter are preferred due to their stability against the lithium metal. These types of batteries cannot be constructed with the previously described type of battery technologies and cell stacking. On the contrary, these type of battery require a new cell stacking architecture to ensure a sufficient supply of oxygen to the cathode. Currently, lithium air batteries have showcased specific energy of 250 to 300 Wh/kg. But this type of battery has the potential for achieving promising theoretical cell level specific energy of 3000 to 3500 Wh/kg. Researchers argue that the lithium air battery is expected to surpass both the lithium ion and lithium-sulfur technology after 2030. It has been found that this technology could achieve 600 to 750 Wh/kg within the next 15 years and 1200 to 1400 Wh/kg within the next 30 years. This specific power of the batteries is expected to improve by a special coating of the electrodes.

Challenges

Even though lithium air batteries show huge potential in their specific energy, they are also subject to a lot of challenges. One of the challenges for lithium air batteries is that due to the presence of atmospheric gases in the air, some of the unwanted chemical products, such as LI_3N, LI_2CO_3 or LIOH may form in the cathode that affects the cathode negatively in its cycle life. Furthermore, one of the challenges for this type of battery is that as the battery discharges the oxygen particles deposit over the cathode which causes potential weight gain over time. (Placke et al. 2017; Sahoo et al. 2020; Duffner et al. 2021; Mauler et al. 2021). This could particularly be a problem in aviation as an increase in weight during flight would increase costs and reduce the distance of travel. The LAB battery has also currently shown a limited number of charging and discharging cycles. During different forms of research and testing, it has also been found that high lithium metal reactivity exists in these types of batteries and gives rise to safety issues, such as overvoltages, especially during charging (Placke et al. 2017; Sahoo et al. 2020; Duffner et al. 2021; Mauler et al. 2021). The realization of LAB battery technology in the near future may be impractical due to the need for fundamental changes in electrolyte components or the membrane.

2.7.2.4 Lithium and Sulfur Availability

Even if batteries with high specific energy density are available by 2035, there would be a need for enough lithium availability to serve the demand from multiple sectors, since it is one of the major resources required to build batteries. The abundance of lithium in the crust is estimated to be between 17 and 60 mg/kg making it the 32nd most abundant element (Tabelin et al. 2021). It has been found that the global lithium reserves in the year 2016 were 14 million tonnes and global lithium resources were 46.9 million tonnes. Ten years before, in 2006, the lithium reserves value was only 4.1 million tonnes. Furthermore, if lithium sulpur batteries reach a mass level of production then also there will not be resource shortages in the system. This is because Almost 78000 thousand metric tons of Sulpur was available as a resource on Earth in 2020 making it the 32nd most available element on Earth (USGS 2021).

This increase in the reserves indicates that despite the remarkable growth of the demand for battery production there is no risk of the exhaustion of natural reserves even in the long term. This remains true even when various other uses of lithium are taken into account (Öko-Institut e.V. 2018). One of the aspects that could slow down battery manufacturing processes could be the slow development of lithium extraction plants. Furthermore, authors like (Von Drachenfels et al. 2021) have tried to analyze the possibility of scale-up of battery cell manufacturing and found that some of the aspects can be scaled easily while others, such as the energy requirements, need more elaborate modeling methodologies. On the other hand, it is expected that the battery recycling industry will develop quickly and 10% of lithium demand can be satisfied by secondary material from electric vehicles (i.e., recycled lithium-ion batteries) by 2030. This figure is expected to rise to around 40% by 2050, although it must be noted that in order to reach these ambitious recycling targets, it will be essential to ensure that batteries are collected and recycled as efficiently as possible (Öko-Institut e.V. 2018). Overall, in the studies mentioned above, certain possibilities of electric aircraft have been assessed, but it is unclear if it will be possible as research in this field is ongoing. In this context, electric aircraft might not become a reality, but if they do, this thesis provides insight into the requirements for them to become operational and their potential role in emission reduction.

2.7.3 Hydrogen Technology

Hydrogen powered aviation is one technology that has gained high attention from the industry. This is mainly because of its emissions reduction potentials compared to conventional kerosene based propulsion. Recently, airbus has pledged to develop the world's first hydrogen powered zero-emission commercial aircraft by 2035 (Airbus 2021). Another company named ZeroAvia is aiming to build a regional commercial hydrogen-electric aircraft by 2024 (Nicolay et al. 2021). But research related to the application of hydrogen in aeronautical space has been done since the late 20th century and early 21st century. This included projects such as Suntan (USA-1956), Hy-Shot (Australia- 2001), NASA X-43 (USA- 2004), Tupolev Tu-155 (Soviet Union- 1988), Phantom Eye (USA- 2013), Cryoplane (Europe- 2000). Specifically, airbus and NASA have done research into cryoplanes in the past that can have hydrogen as fuel in aircraft with a tank in the fuselage (Verstraete 2013; Baroutaji et al. 2019). From the current level of research, it is found that there are two ways in which hydrogen can be used in aircraft. First, through thermal conversion in H_2 combustion engines and second, through electro-chemical conversion in fuel-cells. Since the integration of H_2 storage on board would be needed, the need for a new aircraft design would be imperative. Due to major differences in the energy density and power systems, H_2 combustion can potentially be used for medium and long-range aircraft, on the other hand, H_2 fuel-cell might be used in the short-haul range segment (Fuel Cells and Hydrogen Joint Undertaking 2020; Hoelzen et al. 2022b).

2.7.3.1 Technical Overview and Requirements

From a technical point of view, weight is of particular importance in aviation. Hydrogen can potentially be used as aircraft fuel as its energy density is 2.6 times compared to jet fuel (Yilmaz et al. 2012; Nicolay et al. 2021; IATA 2023). However, in aviation, the volumetric energy density of aircraft fuel holds significant importance. The maximum volumetric energy density of hydrogen is achieved through liquefaction, a process involving cooling the hydrogen to 21.15 K or—253°C. To store liquid hydrogen (LH_2), the aircraft would require a special cooling system for its storage and a volume 4 times bigger than kerosene to store the same amount of energy. Furthermore, for the storage of LH_2, the tanks must have a spherical or cylindrical shape to allow for insulation requirements and a small differential pressure and have highly efficient vacuum insulation to maintain the liquid temperature for several days (Yilmaz et al. 2012; Nicolay et al. 2021). This is one of the reasons for the research on cryogenic planes in the early 2000s. The aim was to explore the different options to store the hydrogen tanks in the fuselage. Analysis suggests that cryogenic hydrogen tanks, despite requiring more space, are highly competitive with kerosene tanks in terms of overall mass. This advantage is further increased if a fuel-cell system is used. It is found that to achieve the best results, cryogenic hydrogen tanks ought to be tailored to a specific application, including new aircraft designs, as well as flight missions. This is because the insulation design depends on the overall mission profile (Winnefeld et al. 2018). Research has shown that aluminum shows minimal susceptibility to hydrogen embrittlement and this makes it a prime candidate for LH_2 tanks for aircraft. Furthermore, because of their low-density and high strength, aluminum composites offer a potential weight saving of 25% on the tank wall, in comparison to the latest aluminum tanks (Yilmaz et al. 2012). With all the aspects concerned it is believed that for a large commercial aircraft, hydrogen can potentially be used as a fuel. Although, integrating it can cause an increase in drag, which causes higher energy consumption by 9–14% (Verstraete 2013). Also, the structure of the storage tank can increase the OWE (operating weight empty) by 23%. On the other hand, depending on the mission of the aircraft it was found that due to the high energy by weight ratio of hydrogen, there was a high amount of variation in the MTOW (Maximum take of weight) in the interval between + 4.4% and − 14.8% (Godula-Jopek and Westenberger 2016).

The use of fuel-cell technology for aircraft propulsion still is in its nascent stage but recently some other advantageous aspects have been found. Airbus and Boeing are presently investigating the potential use of fuel-cell technology to drive the Auxiliary Power Units (APUs) of their aircraft, which were traditionally powered by diesel engines. Fuel-cells exhibit promise as an alternative to batteries for powering various devices and systems within aircraft. For example, they can be applied for cockpit instrumentation, environmental control systems, lighting, cabin pressurization, actuation of control surfaces, wing anti-ice protection, and landing gear.

There are many other uses of the fuel-cell hydrogen based systems in the aircraft. Among various other technologies, proton exchange membranes (PEM) which are purely fueled with hydrogen and solid oxide fuel-cells SOFC are one of the two most commonly explored applications for aviation. One of the advantages of the PEM type of application is that it requires pure hydrogen without any amalgamation required as the fuel operates at a temperature range between 80–120°C. Furthermore, the PEM cell has a battery specific energy of 600–800 Wh/kg, which makes it quite attractive for aerospace applications (Sahoo et al. 2020). On the other hand, the SOFC is operated with different types of fuel sources and requires an operating temperature between 750–1000°C. One of the applications of hydrogen fuel-cell technology is in using it in a hybrid setting. This means that the hydrogen fuel-cells will work in combination with the gas turbines to improve efficiency and reduce core noise. Boeing has recently developed various propulsion concepts that are based on a combination of fuel-cell/battery and fuel-cell/gas turbines in order to reduce fuel burn and emissions. It has also been found that the German Aerospace Research Center DLR is also developing a purely hydrogen, fueled PEM technology based 4 seater aircraft named HY4 with 80 kW capacity (Sahoo et al. 2020).

2.7.3.2 Emission Reduction Potential of Hydrogen Concepts

In terms of emissions reduction, research shows that hydrogen as fuel or using it in the fuel-cell produces no CO_2 emissions and the only other emissions are NO_x and water vapor. Specifically, hydrogen combustion produces 2.6–4.3 times more water vapor and 80–75% less NO_x emissions than kerosene fuel (Ponater et al. 2006; Khan et al. 2022). Even though the radiative forcing of water vapor forming contrails is lower than that of CO_2 it still causes global warming as discussed in Sect. 2.5. (Ponater et al. 2006) assessed the isolated and cumulative impacts of emissions from a hydrogen-based flight compared to a kerosene-based flight. The study determined that the beneficial effects of zero-CO_2 combustion would counterbalance the disadvantage of increased water vapor exhaust. Additionally, CO_2 has an atmospheric lifetime of up to 100 years, whereas water vapor's lifespan ranges from a few days to 1 year. However, it must be noted that the effects of emissions on ozone formation and cirrus induced clouds will need careful analysis and more research into the radiative forcing of hydrogen combustion by-products to increase the certainty of these predictions (Nicolay et al. 2021; Khan et al. 2022; IATA 2023).

2.7.3.3 Challenges for Hydrogen Propulsion Concepts

There are many challenges to using LH_2 as fuel in the aircraft, for instance, the problem regarding boil-off in storing LH_2 in tanks. In spite of efforts for insulation, ambient heat input cannot be fully avoided while storing liquid hydrogen and therefore, the pressure in the tanks is increased over time. To prevent tank damage due to overpressure, it must be equipped with a venting valve. This means that evaporation accounts for a loss of about 0.3–3% hydrogen per day (Nicolay et al. 2021). But to avoid this problem, thicker insulation can be installed to reduce evaporation losses, resulting in a larger tank system mass and volume. However, it is believed that in the aviation industry, this type of loss is tolerable because most aircraft are usually continuously operating without having long turnaround times. Even then as the airline business relies highly on costs, the increasing amount of boil-off can be a hurdle for the airline companies. This is especially true if hydrogen is available at US $1.6–3.03 per kilogram which is more than 2 times higher than kerosene. For hydrogen to be adopted on a mass scale this problem would surely need to be addressed in the future (Janić 2014). Another problem with using hydrogen in aircraft is that the energy density of liquid hydrogen is only about a quarter of that of jet fuel. This means that for the same amount of energy to be carried on board it needs a storage tank four times the size. The challenge is mitigated with the adoption of fuel-cell technology due to its high efficiency, leading to a further decrease in the fuel load. Fuel-cells offer additional benefits for electric aircraft, such as distributed propulsion and enhancing aerodynamic efficiency (Kadyk et al. 2018). However, as a consequence, aircraft may need to either accommodate fewer passengers to allocate space for storage tanks or undergo a significant increase in size. The former option, applicable to Airbus's initial two concept planes, would result in a reduction in ticket revenue all else being equal. The latter option, exemplified in Airbus's third concept, necessitates a larger airframe, susceptible to increased drag (Airbus 2021; BBC 2021; Nicolay et al. 2021). Both of these problems create hurdles for hydrogen to be adopted in the air transport industry. Especially in the first case, since airlines are highly competitive, a decrease in revenue would have to be traded with a reduction in flight frequency to balance the two things. From the production side, there is a question of mass production of hydrogen at a competitive price in an environmentally friendly way. For instance, global demand for pure hydrogen has been increasing by roughly 4% since 1975. This demand is only expected to increase as hydrogen will be required by many industries and not just the aviation industry. In 2020, 87 million tons of hydrogen was produced which resulted in 900 million tons of CO_2 emissions (IEA 2021a). The high amount of emissions from hydrogen production was mainly because most of the hydrogen produced today is through fossil fuel methods (steam methane reforming SMR) and only 1% is produced through electrolysis. In the research conducted by the International Energy Agency (IEA), it is found that in the net zero scenarios, by 2030 total hydrogen production could be increased to 100–200 million tons and out of this 8–80 million tons could be green hydrogen. By 2050, about 250–500 million tons of hydrogen could be produced and 120–250 million tons could be green hydrogen. These potential production levels could be not enough when aligned with the potential aviation needs of the future. For instance, reports and expert opinions suggest that between 2035 and 2040 about 8–16% of the total global fleet could be replaced by hydrogen aircraft. Furthermore, by 2050 40–60% of the global fleet could be replaced by hydrogen aircraft. For this transition, from 2035 until 2050 the aviation system could require 5–130 million tons of hydrogen per year (IEA 2019, 2021a, b; Fuel Cells and Hydrogen Joint Undertaking

2020). Whether it is green or fossil fuel based hydrogen that can be available for aviation and that too at a competitive price remains uncertain. Another aspect of using hydrogen in aviation is the infrastructure required for it to be used at airports. If the hydrogen is not produced at the airport, then alternatively it can be in its liquid or gaseous form transported to the airport. Transport modes for hydrogen could be trucks, rail, vessels or pipeline systems. If the hydrogen is produced on-site then transportation would not be needed but in either case, the hydrogen delivered at the airport would need large tanks for it to be stored. These storage systems would need to have the ability to keep hydrogen in liquid form. Since hydrogen is liquid at—253°C the liquefaction system might require a lot of energy (Nicolay et al. 2021). The refueling of liquid hydrogen to the planes can be carried out through two methods refueling trucks or refueling pipelines and hydrant systems. The use of refueling trucks enables a more flexible deployment but these can contribute to increasing traffic on airport aprons. Also, the loading volume of one refueling truck is limited and this could be not enough for large airports and aircraft. Due to these reasons, it is believed that using a hydrogen pipeline and hydrant system could be a better alternative. It is found in the literature that these systems could be built underneath the airport apron and accessed through a hydrant at each aircraft stand (Hoelzen et al. 2022a). To build these capacities in the future at airports, significant costs would be incurred. As refueling infrastructure and operating costs are typically integrated into the fuel price paid by the aircraft operator rather than a distinct fee to the airport, this setup can result in an elevation of operating costs for airlines. Research indicates that at airports with less than 125 kt LH_2 annual demand, an LH_2 refueling truck setup proves to be the more economically viable option. Conversely, for airports with higher annual LH_2 demands, an LH_2 pipeline and hydrant system may yield slight cost reductions and facilitate safer and quicker refueling processes. However, the refueling system was found to only cost 3 to 4% of the total supply costs of LH_2 in all demand scenarios. The costs are dominated by the costs for green H_2 produced offsite followed by the costs for liquefaction of H_2 at an airport (Mangold et al. 2022; Hoelzen et al. 2022b, a).

2.8 Requirements for Analysis: Methodological Design and Input Parameters

After discussing the components of the air transport system in detail, it is important to lay down the requirements that are necessary to answer the research questions posed for this thesis. As mentioned in Sect. 1.2, the primary research question for this thesis is to find and analyze the different transition pathways that can help the air transport system move towards a low emissions and sustainable system in the future. To answer this type of problem, several aspects would be required from a methodological and input parameter standpoint.

2.8.1 Methodological Design Requirements

From a methodological viewpoint, Table 2.8 shows a short overview of the requirements in a tabular format. First, to answer the research question a dynamic modeling approach would be required because the airline industry is highly cyclical in nature and involves a high amount of dynamic interaction with different components in the system (Lyneis 1999). Second, the airline industry is dominated by feedback loops that generate industry trends and only dynamic models can capture the feedback loops in the system more efficiently to replicate the industry trends. Furthermore, dynamic models also provide a better understanding of the causes of the industry behaviors and with the help of this, an early detection of changes in structure can be perceived beforehand. This helps in accurately depicting the real world and its dynamic changes overtime. Accurately capturing the real world behavior is imperative to conduct this analysis because most climate objectives laid down for the aviation system are set for a medium to long term time frame and span until 2050. Therefore, the methodology that can provide accurate depiction of the real world can help in testing different scenarios. Third, for this study, assessment from a continuous time period perspective is a major requirement instead of singe discrete event analysis approach. The main reason for this is that the strategies needed to be tested not only require an end result but also an assessment of their behavior over time. The assessment of the behavior of the system over time provides clarity on the strengths and flaws of each strategy.

Furthermore, the assessment of variables over time is also important because the behavior of the system during the transition period i.e. 2025–2035 is extremely important. The main reason for this is that it helps in choosing the most sustainable strategy of all other strategies as certain strategies can have good end results but show unrealistic aspects during the transition periods which can be dangerous or unattainable for the system. Fourth, due to the nature of the research question the methodology chosen for the study should be able to provide a possibility to consider a high level of aggregation in variables.

Table 2.8 Methodological design requirements to conduct the analysis

Methodology Design	Requirements
Structure	Dynamic
Basic building blocks	Feedback loops
Handling of time	Continuous
Level of design	Aggregated at macro level

The main reason for this requirement is that to assess a system from a high level viewpoint as planned for this study, the variables of interest cannot be considered accurately but have to be considered in an aggregate way. This sort of setting allows to include large number of variables that are relevant to the research question from air transport system sub-sections. Contrary to this, it is not possible to include large number of variables in a discrete event analysis as that approach requires accurate rather than aggregated variables along with rigid modeling structure. Therefore, to conduct this analysis at a high level, a large number of variables with aggregation assessed dynamically would be the best method for this analysis.

2.8.2 Input Parameter Requirements

From an input parameter viewpoint, on an overall level, the methodology for this study should be able to provide the possibility to test a robust set of scenarios using different values for different input parameters at once. Specifically, here robustness refers to the possibility of changing the values of input parameters to test for sensitivity effects on the feedback mechanisms of the system. This is highly necessary to conduct the analysis because the airline industry is highly cyclical and interconnected industry where a change in one sub-section can cause cascading effects across the system. In order to clarify the initial set of possibilities to the stakeholders and the decision-makers of the system, a methodology that can produce an exhaustive set of results based on changes in the input parameter settings at once would be important. This is because testing the scenarios in this way would provide a wide range of confidence intervals that provide clarity on uncertainty in the system in the near future. Also, conducting the analysis in this way would provide the possibility of testing multiple strategies and policies in the system.

In order to realistically test the strategies and policies in the system, a number of aggregated or averaged input parameters related to the different sub-sections of the system would be required. Table 2.9 shows the overview of the parameter requirement in a tabular format. The discussion related to the sub-section wise parameter requirement in accordance with the research questions is as follows:

Airline industry

For the airline industry, the parameters related to the fleet planning mechanism in accordance with the change in demand would have to be included. This is because in order to address the question of the adoption of new technologies, delays in the adjustment of fleet planning, airline ordering, operation mechanism and retirement would be highly necessary. The airline industry is dominated by delivery delays and cycles, therefore, these delays have to be taken into account while addressing problems related to adoption. Also, the different pathways of phasing in of new technology and phasing out of the old technology highly depend on the delay parameters and these scenarios are of high interest for the research question therefore they are a vital requirement. In order to assess the adoption of new technologies in different range segments, the input parameter data for traffic segregation would be required. The segregated fleet data would provide the necessary share of the short, medium and long-haul flight traffic in the grand scheme of global airline traffic. To answer the amount of emission reduction potential of each type of new technology in the system, the input parameter related to the GHG emissions savings of new technologies for in-flight or life cycle based savings would be required. Lastly, to incorporate the aspect of operational costs, the parameters related to the costs of each type of new technology would be required.

Regulators and policy-makers

For the regulators and policy-makers sub-section, the most important parameters to be included would be the policy parameters. Policy parameters here mean the different numerical values that make a policy more or less stringent. This is particularly important because it will help in assessing the effectiveness of policies and incentives with respect to the type and amount of GHG emissions it can help reduce.

Aircraft manufacturers

The most important parameter requirement for the aircraft manufacturers subsections is the parameters related to the delays in capacity adjustment for the manufacturing of the new technologies. The delays in the capacity adjustments would provide insights into the changes in adoption rates with respect to different policies and strategies tested. Furthermore, it would also provide help in clarifying the changes in the adoption and emissions in the presence of different scenarios related to the demand growth patterns in the industry.

Table 2.9 Overview of input parameter requirements to conduct the research

Sub-section	Input parameter requirement
Airline industry	• Parameters related to fleet planning of new technologies (SAF, Hydrogen and electric aircraft) o Fleet adjustment o Aircraft ordering o Operations and retirements • Traffic segregation • Emission reduction potential of SAF, Hydrogen and electric aircraft • Operating costs related to SAF, Hydrogen and electric aircraft
Regulators and policy-makers	• Parameters related to policies o CO_2 tax o Jet fuel tax o Seat tax o Fleet restriction
Aircraft manufacturers	• Parameters related delays in capacity adjustment for o Hydrogen aircraft o Electric aircraft
Fuel producers	• Parameters related delays in capacity adjustment for o SAF o Hydrogen • Resource availability o Availability of hydrogen o Availability of green electricity o Share of green hydrogen • Emissions related to production of hydrogen
Air travel demand	• Input parameters related to factors that influence demand o GDP per capita o Population o Jet fuel prices

Fuel producers

The parameters requirement for the fuel producer sub-section would be the values related to delays in fuel capacity adjustments, especially related to the SAF as it is believed to be the leader in helping emission reduction for the air transport system. Also, for this sub-section, the parameters related to the initial amount of green electricity availability, the share of green hydrogen, initial amount of hydrogen availability by the time new technologies are introduced in the system would be required. This is because these values would be necessary in computing the electricity and hydrogen requirements due to the adoption of new technologies in the system. Lastly, the parameters related to the amount of emissions emitted by producing hydrogen would also be required as it will help in computing the emissions from the production side.

Air travel demand

For the air travel demand, the parameters related to the aspects that influence the demand would be required. This includes macroeconomic parameters that are discussed in Sect. 2.2. Furthermore, the parameters related to the growth in these parameters would also need to be included.

2.8.3 Research Questions: Air Transport System Sub-sections

Based on the discussions in Sect. 2.8 and 1.2, research questions for each sub-section of the system can be formulated. Table 2.10 showcases the formulated question for each sub-section under the secondary question of the thesis in a tabular format.

As discussed in Sect. 2.7 and 2.6, different type of aircraft propulsion technologies and alternative fuels can be used to help transition the air transport system towards low emissions. Regarding new aircraft propulsion technologies, it is expected that battery-electric aircraft, hydrogen-electric aircraft and hydrogen combustion aircraft can be at the forefront to help in transitioning of the system. For alternative fuels, different pathways to produce SAF, especially, HEFA in recent years and PtL for the future show high potential. In this context, the primary sub-question for the sub-section of the airline industry is which of the new technologies and fuels (SAF, battery-electric propulsion concept, hydrogen combustion and fuel-cell) can play a major role in transitioning the air transport system towards sustainability?

Regarding the Regulators and policy-makers sub-section, in Sect. 2.5, policies like jet fuel tax, carbon seat tax and quota policies that can help achieve sustainability in the air transport system were discussed. In addition, strategies like Flightpath 2050 CO_2 emissions goal and IATA net zero 2050 goal are being planned to achieve low emissions in the near future. The primary question for the regulators and policy-makers sub-section for this thesis is what kind of policies, incentives, quotas and strategies (fleet restriction, quotas, taxes etc.) would be required to be able to achieve emission goals by adoption of new technologies in the system?

For the aircraft manufacturers sub-section, in Sect. 2.7, it was discussed that there are still a lot of challenges that manufacturers face to make hydrogen and battery-electric aircraft a reality. If the challenges are overcome, then one of the most important factors that would affect the adoption of new technology and the reduction of emissions would be delays in production. This is because manufacturers would need to adjust their production capacities dynamically to the fleet demands of the airline industry. Thus, the primary sub-question for the aircraft manufacturer sub-section is how can the aircraft manufacturers adjust their capacities after the introduction of new technologies in the system. Furthermore, it would be essential to know the role of manufacturing delays in the adoption of new technologies.

For the fuel producer's sub-section, in Sect. 2.6 and 2.7.3.3, a discussion was done on the potential of different types of alternative fuels such as drop-in (SAF) and non-drop-in (hydrogen) fuels. Specifically, regarding SAF, in Sect. 2.6.6 the limitation to produce SAF from HEFA and FT-synthesis production pathways due to the limitations in feedstock availability was discussed. Due to the lack of feedstock availability, it is believed that PtL type of SAF would be preferred to meet the alternative fuel demands after 2030. In addition, similar to aircraft manufacturers, the delays in alternative fuel production could also play a vital role in the adoption of SAF in the system. With respect to the capacity adjustment and delays to produce SAF and hydrogen, the primary sub-question for the fuel producer sub-section is how the alternative fuel producers (hydrogen and SAF producers) can adjust their production capacities with respect to the adoption of new technology? Furthermore, how does the change in the fuel production delays and scaling effects affect the transition of the system in terms of the adoption of new propulsion technologies and emissions reduction?

By answering the above questions, the aim of this work is twofold, first, to provide the industry stakeholders about the potential contribution of the new technologies in different range segments in the presence of different policies and strategies to achieve the ambitious goals. Specifically, stakeholders here mean the airline industry, aircraft manufacturers, fuel producers and regulators and policy-makers. For policy-makers, testing different policies in this way would provide the decision-makers with an initial impression of possibilities on a system level and help further in their planning process. For fuel producers, this analysis would also be able to provide the amount of fuel and electricity requirements to meet future demands robustly. By learning the potential requirements in different scenarios considered, the industry fuel producers can start to better prepare themselves to develop their capacities accordingly or pursue other strategies. For the airline industry, with the help of the initial impression of the new technologies the airline companies can start to devise twin fold strategies that would help them to operate the technologies for a profit and reduce in-flight emissions at the same time. For aircraft manufacturers, with the help of this analysis, they would be able to understand changes required in the manufacturing capacity due to the changes in adoption, policy implementation and demand growth patterns.

Second, by conducting this analysis a gap in the existing literature is filled by developing a model that incorporates the most important sub-systems of the aviation system such as aircraft manufacturers, airline companies, air travel demand, aircraft fuel producers, environmental emissions and aviation regulatory authorities. Developing the system level model would open the future prospects of research in the air transport system which allows the researchers to also test for legal framework delays and business strategies of the airlines.

Table 2.10 Research questions for each sub-section of the system

Research Questions: air transport system sub-sections			
What are the different transition pathways through which the air transport system can transition towards a low emissions and sustainable system in the future?			
Secondary Research Question	how would the aviation emissions goals be achieved and what would be the requirements for the same?	Airline industry	• Which of the new technologies (SAF, electric propulsion concept, hydrogen combustion and fuel-cell) can play a major role in transitioning the aviation system towards sustainability? • How would these technologies be adopted in the short, medium and long-haul segments in presence of delays? • How much emissions reduction potential would they provide in these different flight segments?

(continued)

Table 2.10 (continued)

Research Questions: air transport system sub-sections		
What are the different transition pathways through which the air transport system can transition towards a low emissions and sustainable system in the future?		
	Regulators and policy-makers	• Can the new technologies help in achieving the different targets set by the regulators for the aviation industry (Flightpath 2050 CO_2 emissions goal, IATA net zero 2050 etc.)? • If yes, then how much can they help in achieving and if not then what kind of policies, incentives and strategies (fleet restriction, quotas, taxes etc.) would be required for them to be able to be effective? • Which type of emissions (CO_2, NO_x, O_3 and contrails) can the different technologies help in reducing from the environment under different policy settings in different range segments?
	Aircraft manufacturers	• How can the aircraft manufacturers adjust their capacities after the introduction of new technologies in the system? • What would be the role of manufacturing delays in the adoption of new technologies? • What would be the effects of changes in aircraft manufacturing delays on the adoption and emissions over time? • How would the manufacturers adjust their capacities due to the changes caused by the changes in demand patterns in the system (high, current or low growth scenarios)?
	Fuel producers	• How can alternative fuel producers (hydrogen and SAF producers) adjust their production capacities with respect to the adoption of new technology? • How does the change in the fuel production delays and scaling effects affect the transition of the system in terms of adoption of new technology and emissions reduction? • How much hydrogen and green electricity requirements would be needed to produce the required fuel for the system? • How much emissions (CO_2 and CO_2 equivalent) would result from producing the required hydrogen for aviation needs in the future?

Table 3.18 (continued)

3 Literature Review

In this section, the literature regarding the research in the air transport system (ATS) is discussed. Specifically, the system dynamics methodology based literature is discussed in more detail as it is used to answer the research questions of this thesis. The research conducted by other institutions, organizations and action groups in the air transport system is also discussed in this section.

3.1 Review Method

To conduct the literature review, a systematic approach was utilized to include all the literature in the relevant area of the topic of this thesis and to reduce potential biases regarding identifying, selecting and analyzing articles (Booth et al. 2021).

Figure 3.1 shows the step-wise procedure carried out for the database search to find the relevant literature and conduct a structural analysis. In the first step, a database search was carried out in March 2024, using the Scopus database from Elsevier (www.scopus.com). The Scopus database was used because it provides an extensive number of articles from peer-reviewed journals in the scientific community. To find the relevant literature, the first search string in terms of context with the keyword "air transport" along with the second keyword "sustainable" was used. These two keywords were used to search in the Scopus database in the titles, abstracts or keywords of different articles. These two particular keywords were chosen because the topic of the thesis centralizes around the sustainable air transport system of the future. In other words, the literature regarding sustainability of the air transport system was searched. Since this sort of search

C. Talwar, *Transition Pathways towards a Low Emission Air Transport System of the Future – A System Level Analysis*, Produktion und Logistik,
https://doi.org/10.1007/978-3-658-46507-0_3

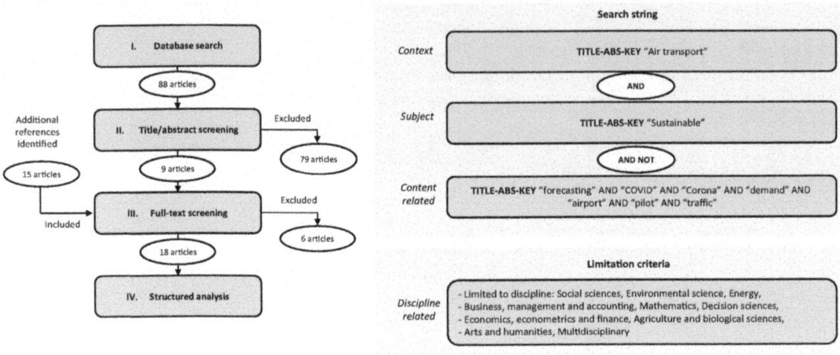

Fig. 3.1 Overview of the systematic search and screening process

can provide a large number of irrelevant literature, content-related omission was done using keywords. Specifically, these keywords were forecasting, COVID, demand, Corona, airport, pilot, and traffic. Excluding articles with these keywords would remove the irrelevant literature from the database search string. Furthermore, for the limitation criteria, the selection of the most relevant disciplines was chosen. These included social sciences, environmental science, energy, business management and accounting, mathematics, economics, econometrics and finance, agriculture and biological sciences, arts and humanities, and multidisciplinary. This limitation was done to exclude articles from disciplines that are not relevant to the goal of this work and the research question.

In the next step, a two-step screening was done for the articles found through database search. This was done to remove the unimportant literature in the database search. The first step of screening was done by analyzing the relevance of the article through title or abstract. This allowed to screen out 79 articles from the total database search of 88 articles. The remaining nine articles were screened by full-text analysis. Through the full-text analysis, it was found that 6 out of 9 articles were irrelevant to the thesis. Within the three important articles, regarding the topic of the thesis 15 other articles were found with the help of snowballing and in-text references. In total 18 articles were found that were relevant to the topic of this thesis. The 18 articles were then analyzed, based on the requirements mentioned in the previous section to conduct the research aim of this thesis. Table 3.1 shows an overview of the consideration of requirements in the literature and the analysis of different articles based on the requirements. Even though the previous studies were robust and insightful, they did not include new technologies like hydrogen and its role in different flight segments and their emissions reduction potentials. Previous models also did not take into account feedback effects on the capacity changes of the fuel producers due to an increase in usage. Furthermore, since certain resources like green electricity are scarce and are required to produce green hydrogen and certain types of SAF, previous studies also did not take into account the feedback effect of an increase in fuel production capacity on required changes in green electricity availability. All these aspects that are missed in the previous models are vital in assessing the transition of the commercial air transport system towards sustainability and cannot be overlooked. From the table, it can be observed that there has been no study in the past that has taken into account all the required considerations. Therefore, there is a major gap in the literature regarding the system level analysis of the air transport system in the presence of policies and new technologies. In this thesis, precisely the missing aspects are added to the model to conduct the analysis and provide insights into how the feedback loops affect the outcome of the system in the simulation model.

3.2 System Dynamics and the Air Transport System

System dynamics (SD) is an approach to understanding the nonlinear behavior of complex systems over time. By using SD, the interactions of various systems present around us can be understood in an interconnected and holistic manner. In system dynamics, the idea of circular feedback is used to explain the feedback effects in a system. System dynamics has been used in a vast variety of disciplines, such as mathematics, biology, ecology physics, history, and literature (Sterman 2000).

To explain briefly, system dynamics modeling is a discipline for understanding systems as a whole. By using system dynamics, researchers can aim towards developing a framework for observing the interaction of the interrelationships of different factors rather than for factors themselves. This sort of modeling helps in understanding patterns of change rather than static snapshots (Sterman 2000). System dynamics modeling is derived from systems thinking, which is the discipline of observing the structures that underline the complex behaviors to understand the broader system. There have been researchers in the past who have applied the use of system dynamics for problems related to aviation industries. For instance, (Quan 1997) utilized the SD methodology to forecast the demand for the transportation system in China from two dimensions by modeling China's microscopic and macroscopic social economic development. In their modeling, the potential air travel demand was referred to as a function of some social economic variables involving aspects like industry, society, population and infrastructure that determine the demand. It consisted of airlines, airports, and air routes to determine the demand. This model was mainly developed to provide suggestions for policies in the development of China's transportation sector. Moreover, this model provided some quantitative suggestions for such policies.

(Suryani et al. 2010) applied a System Dynamics (SD) model to project the prospective air passenger demand at Taiwan airport. This research was done in order to understand the qualitative and quantitative requirements of the runway and passenger terminal expansions. In their model, they included several feedback mechanisms and validated their model by testing it against the historical data. Their results indicated that airfare, level of service, GDP, population, number of flights per day and dwell time play an important role in determining the air passenger volume and demand. Their results helped in understanding the key differentiation between different types of demand scenarios and the requirements for the same expansion of the runway and passenger terminal.

There have been some authors who have used the SD model to answer the questions from airline industry in the area of airline entry exits, and alliance. (Kleer et al. 2008) utilized the SD model to analyze the impact of airline entry on national routes under different scenarios. Their model consists of several aspects, such as stock and flow theory, policy, attractiveness, airline configuration, and demand. The authors calibrated their model by using the monetary information of the German antitrust lawsuit between Lufthansa and Germania. The authors use this data to simulate several strategic movements of airlines. Using the model, the authors were able to confirm the hypothesis that dynamics in the airline market were driven partially by the effects of airline entry and exit on a city. Furthermore, in their paper, the authors also provided suggestions for carriers to optimize their cost structure and impede the entry of other carriers on the same route.

(Agusdinata and De Klein 2002) use the causal loop diagram (CLD) a technique, often used in system dynamics methodology to detect the driving forces for airline alliances. The authors used these causal loop diagrams to determine and examine the stability of alliances under a global economy scenario. The causal loop diagrams demonstrated the complex behavior within the economic cycle. In their paper, the authors describe the relationship between the transition economy and airline alliances. They inferred from their findings that globalization not only creates a competitive market environment for airlines but also generates opportunities. Their analysis consisted of external stability drivers, such as the cyclical economy and anti-trust legislation and hub airport congestions. In terms

Table 3.1 Overview of the considerations of requirements in the literature

Article	Requirements: Considerations in the literature							
	Consideration of cyclical dynamics	Policy analysis	Consideration of new propulsion concepts	Consideration of biofuels/ SAF	Demand projection and shifts	Consideration of resources for SAFs	Consideration of inflight emissions	Consideration of production side emissions for SAF
Quan 1997	◐	●	○	○	●	○	○	○
Suryani et al. 2010	●	●	○	○	●	○	○	○
Kleer et al. 2008	●	●	○	○	●	○	○	○
Agusdinata and De Klein 2002	◐	○	○	○	●	○	○	○
Bießlich et al. 2014	○	○	○	○	◐	○	○	○
Lyneis 2000	●	○	○	○	●	○	○	○
Liehr et al. 2001	●	○	○	○	●	○	○	○
Pierson and Sterman 2013	●	○	○	○	●	○	○	○
Kieckhäfer et al. 2018	●	●	○	●	●	○	●	○

(continued)

Table 3.1 (continued)

Article	Requirements: Considerations in the literature							
	Consideration of cyclical dynamics	Policy analysis	Consideration of new propulsion concepts	Consideration of biofuels/ SAF	Demand projection and shifts	Consideration of resources for SAFs	Consideration of inflight emissions	Consideration of production side emissions for SAF
Sgouridis et al. 2011	●	●	○	●	●	○	●	○
Csala and Sgouridis 2013	●	○	○	○	◐	○	◐	○
Eurocontrol 2022a	○	●	●	●	●	○	●	○
Ghosh et al. 2017	○	○	○	○	◐	○	○	○
Abrantes et al. 2021	○	○	◐	●	◐	○	●	○
Abrantes et al. 2024	◐	○	◐	○	◐	○	●	○

(continued)

Table 3.1 (continued)

Article	Requirements: Considerations in the literature							
	Consideration of cyclical dynamics	Policy analysis	Consideration of new propulsion concepts	Consideration of biofuels/ SAF	Demand projection and shifts	Consideration of resources for SAFs	Consideration of inflight emissions	Consideration of production side emissions for SAF
O'Connell et al. 2019	○	○	○	●	○	●	○	●
Kousoulidou et al. 2016	○	○	○	●	◐	○	●	○
Åkerman, J., 2005	○	○	◐	○	◐	○	●	○

● Completely fulfilled ◐ = Partly fulfilled ○ = Not fulfilled of internal stability factors, the authors assessed factors like alliance organization and inter organizational relationships. This paper provided a foundation for quantitative analysis and sophisticated models for studying the changes in global alliances and behaviors of airlines in the changing economic environment.

(Bießlich et al. 2014) used the SD methodology to build a generic model involving elements from both operational and economic areas. The authors incorporated data from Hamburg airport to assess the functionality of their model. Their model proved to be a highly useful tool to showcase the dynamics of aircraft and customer movements, number of passengers, capacity requirements, infrastructure development, and cash flow. The primary focus of their model was the development of the long-term plan of an airport from an infrastructure planning and development perspective.

The studies that are specifically related to the questions presented in this thesis included the analysis of airline cycles, policies to curve the CO_2 emissions in the system, and the introduction or adoption of biofuels in the system.

In this regard (Lyneis 2000) used the system dynamics model for the commercial jet aircraft industry to demonstrate its effectiveness in forming reliable forecasts for the industry. The aim of this model was to develop cast and showcase that the system dynamics model can display the industry trends that have been going on for the past few decades. The article stressed three main points.

(1) System dynamics model could provide realistically a more reliable forecast of short to medium trend statistical models which in turn positions the decision-makers to make better decisions.
(2) (Lyneis 2000) argues that system dynamics model provides a means of understanding of the causes of industry behavior. With the help of this, the authors, believe that the early detection of changes in the industry structure, and the factors which forecast behavior can be done.
(3) System dynamics models enable the identification of plausible scenarios to inform decision-making and policies.

With their model, the authors were able to illustrate the important structural changes required in the industry and suggestions on how to avoid unnecessary capacity expansions for the best business outcomes.

(Liehr et al. 2001) used a SD model to identify the key factors for cyclical behavior and poor returns on the invested capital in the airline industry. Their model consisted of a generic cycle that generated a feedback structure of the problem with negative feedback and two delays. Even though this model was simplistic in nature, yet it provided a first explanation for the business cycles in the airline industry. Specifically, the model helped in three different ways. First, it helped to identify the cycle generating components of the industry and further to understand their interactions with other aspects. Second, it helped to analyze the different scenarios. Third, it helped in identifying the key variables and leverages for the cycle management strategies.

The model developed by (Liehr et al. 2001) reproduced the historical behavior of the airline market and helped in the estimation of the future trends for the commercial aircraft industry. Specifically, the purpose of the model was threefold. First, the authors wanted to gain a deeper understanding of the dynamics of cyclical movements and to detect the central structure of the problem. Second, the authors wanted to test alternative policies in order to infer strategies for long-term capacity and fleet planning. Third, to develop a model that could simulate different scenarios, for example, shocks in demand.

The authors inferred that the airline industry's market cycle roughly lasted from 8 to 10 years, which was typically called juglar waves. The jular cycles would typically consists of four phases of economic cycles namely; prosperity, crisis, liquidation and recession. It was believed the juglar waves represented the machine investment cycles and were considered the economic cycle. This sort of research contributed to the widespread opinion of managers that the cycles in the airline industry are responses to fluctuations in the GDP. It was found that as a consequence of this, there existed a lack of cyclical management strategies to smoothen the oscillations and to reduce their negative impact on profitability. Contrary to the previous belief, the results of the (Liehr et al. 2001) model suggested that the cyclical behavior observed in the industry is endogenously generated. Furthermore, the authors found that the business cycles were not caused by fluctuations in the world GDP.

The previous model was then extended by (Pierson and Sterman 2013) to explore the causes of airline industry cycles. The authors developed a dynamic model of the airline industry by expanding and including factors such as capacity expansion, demand, pricing, wages, costs, and profit. The authors implemented the model to estimate standard errors in the system dynamics models and highlight important limitations in the current approaches to calibrating larger models. But the main research question of the paper was to find out the aspects that affect profit cycles in the airline industry. Figure 3.2 here shows the graphical depiction of the feedback structure of the model developed by (Pierson and Sterman 2013). The result of their model suggested that price setting feedback plays a surprisingly important role in determining the industry profits stability. Specifically, yield management strengthens a negative feedback loop that stabilizes the industry cycle in capacity, revenue, and profit.

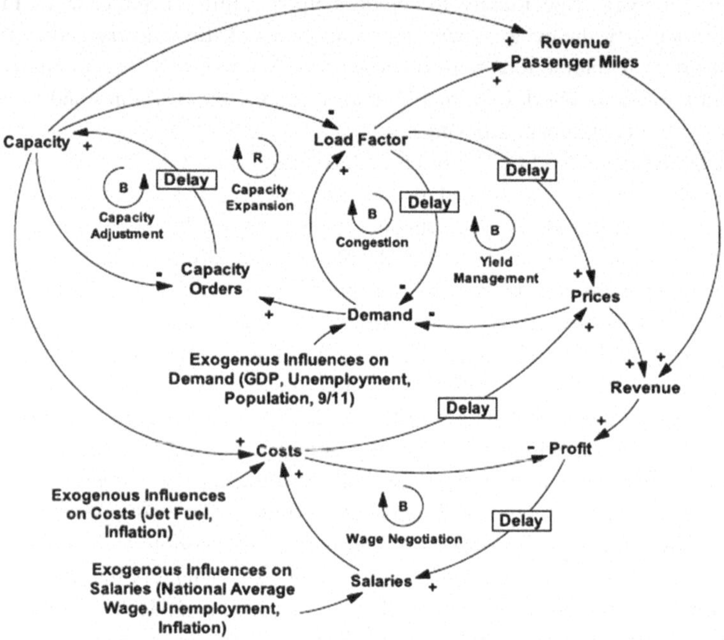

Fig. 3.2 Overview of feedback structure of model build by (Pierson and Sterman 2013)

The use of yield management enhances the responsiveness of prices to demand fluctuations compared to capacity and improves average load factors. Nevertheless, sensitivity tests altering the potency of yield management feedback indicate that, overall, decision rules in airline pricing contribute to increased operational leverage and profitability variance (Pierson and Sterman 2013). This dynamic may position the industry in close proximity to a global minimum in the risk-return spectrum. One of the limitations of their work was that the model was not robust enough to address the important dynamics that arise from different pricing rules at the firm level.

One of the limitations of their work was that the model was not robust enough to address the important dynamics that arise from different pricing rules at the firm level. The above model was then extended by (Kieckhäfer et al. 2018), to study the potential of reducing CO_2 emissions with the help of alternative fuels in aviation. Figure 3.3 shows the new aspects that were introduced in the model such as drop-in quota, alternative fuel and adoption mechanisms. The main aim

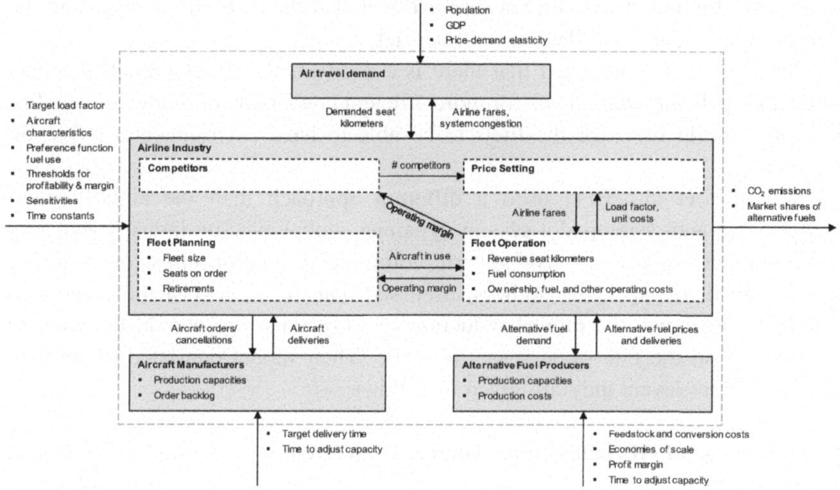

Fig. 3.3 Overview of the system dynamics ATS simulation model developed by (Kieckhäfer et al. 2018)

of the authors was to analyze the possible adoption rates of alternative jet fuels and how they can contribute to aviation emissions reduction. Another aim of the authors to conduct the study was to gain a deeper understanding of the relevant endogenous system behavior and the important variables that account for it. According to (Kieckhäfer et al. 2018), it would facilitate the development of suitable policy measures to accompany the introduction of alternative fuels. Their model consisted of the main actors of the ATS such as air travel demand, airlines, aircraft manufacturers, and alternative fuel producers. The authors incorporated a lot of input parameters which were used to address the uncertainty of the future scenarios. The authors conducted a Monte Carlo simulation analysis to account for these uncertainties. The results of their analysis suggested that the adoption of alternative fuels is rather unlikely for the currently certified production pathways of alternative jet fuels. The main reason for this is that the current production pathways do not allow for competitive prices as compared to conventional jet fuel. On the other hand, to increase the adoption rate, the authors found that the prices of conventional jet fuel would have to experience a high growth rate. Furthermore, an increase in production capacity and monetary incentives would be highly necessary for the higher adoption of alternative jet fuels. It was also found that even if alternative jet fuels are adopted, their long-term climate reduction benefits are limited. The main reason for this is that the restrictions regarding the blend of these fuels with the conventional jet.

The authors also indicated that there is a need for the development of a new production pathway that allows for more efficient conversion of biomass or hydrogen into drop-in fuels for the fuels to be able to have environmental reduction potential.

(Sgouridis et al. 2011) used a different approach than the authors mentioned above and developed a system dynamics global aviation industry dynamic (GAID) model to test for five generic commercial aviation emission reducing policies. In their paper, the authors discussed that, in order to achieve the sustainability objective, the rate of reduction of CO_2 emissions should be equal to or greater than the rate of increase in traffic. Their model consisted of the five most important levers they are given as follows:

- **Technological efficiency improvements**: This refers to the fuel efficiency of the aircraft.
- **Operational efficiency improvements:** This lever included effects of airline operations like aircraft weight reductions and air traffic control operations i.e., fuel optimized flightpath, altitude, reduced ATC delays etc.
- **Use of alternative fuels:** This included the use of alternative fuels that have a lower CO_2 emissions content as compared to jet fuels.
- **Demand shift:** This accounted for the changes in travelers' mode choice behavior or reduction of demand due to alternatives such as video-conferencing, virtual meetings, etc.
- **Carbon pricing (i.e. market-based incentives):** This involved implementing a strategy to raise the effective fuel price and decrease demand by leveraging the relationship between price and demand elasticity.

The authors used these key levers as a basis for generating different scenarios in the GAID model to evaluate the potential CO_2 emissions reduction for the aviation industry from 2004 to 2024. For their analysis, the authors conducted a set of 34 scenarios with various levels of aggressiveness along with five generic policies.

The results of the analysis suggested that the efficiency improvement policies would not generate sufficient emission reduction on their own even if the boundary of the expectations is pushed. In other words, no single policy could maintain emission level study while increasing the projected demand for air transportation. The results suggested that it was partly due to the induced demand that was created by the reduction of the operating cost that negatively affected the average air fares and consequently stimulated demand. Specifically, it was found that a 2% annual improvement in the fuel consumption of new aircraft would only yield less than a 4% reduction in cumulative CO_2 emissions between 2004 and 2024. Similar to this, the operational efficiency improvements of 12% when applied from the year 2008 could only reduce emissions by 7%. Therefore, from this, it was concluded that the increase in technological and operational efficiency caused an increase in induced demand, and therefore reduced the effectiveness of CO_2 emissions reductions.

It was also found that the demand management schemes only became effective if aggressive measures were implemented. In other words, carbon pricing would need to be very high to achieve a reduction in demand. Specifically, it was found that a price of $200 per metric ton of CO_2 would be needed for a total reduction of 8% of emissions as compared to the baseline scenario. In terms of diversion of demand, it was found that a reduction in short-haul flights by 60% by 2024 would be needed to reduce airline emissions by 20%.

In their discussions, the authors mentioned that the simulation results suggest that a combination of proposed policies produces something called a weak sustainability effect. A weak sustainability effect means that an increasing supply to meet new demand happens while maintaining constant or increasing slightly emissions levels. In this regard, a combination of policies that included higher levels of technological and operational efficiencies along with the use of biofuels with moderate levels of carbon pricing could achieve a 20% reduction in emissions as compared to 2004 values in 2024.

The above model was then extended by (Csala and Sgouridis 2013) to explore the dynamic behavior of the competitive aviation industry under different transition scenarios. The authors build a model called "a Global Aviation Industry Dynamic Transition (GAIDT) model" to test for these scenarios. The authors included multiple dynamic behaviors of the airline companies in the model. These were as follows:

(1) Existence of first mover advantage.
(2) Benefit of regulatory mandates for certain transition options.
(3) Effect of the diffusion rate of transition options on the financials of the industry.
(4) Robustness of transition options in the presence of different exogenous parameters.

The aim of the paper was to develop appropriate policies and competitive strategies to facilitate sustainable transition in the air transport system. The strategy assessment was done with the help of sensitivity analysis in the GAIDT model (Csala and Sgouridis 2013). The assessment was separately carried out for one dimensional and multidimensional changes with respect to narrow and wide body aircraft. In all their strategy cases, the authors compared the results to the business as usual (BAU) strategy where the aircraft manufacturers continue with their development strategies of current technology and release aircraft in the planned timeframes, under different fuel price scenarios. The authors also considered three different fuel pricing scenarios as high, current and low growth scenarios to add robustness to the analysis. The different strategies that were adopted in the model by airlines were as follows:

(1) **Innovate:** A situation when one aircraft manufacturer decides to radically change the design of the aircraft, while the other manufacturers develop the currently existing technologies, under different fuel scenarios.
(2) **Follow:** In this scenario, most of the manufacturers decide to fundamentally change the aircraft type and design. This is done by introducing new technologies at different points in time and different fuel price scenarios.
(3) **Industry reform:** In this strategy scenario, most manufacturers decide to completely change their aircraft that can offer substantial fuel burn benefits and introduce it at the same time under different fuel price scenarios.

The results of the analysis suggested that the strategies of the industry would highly depend on fuel prices. It was also found that if fuel prices increase then an innovation strategy would be highly beneficial. When the results were decoupled from the fuel prices, it was found that the late transition options of innovation in the year 2030 would be preferred (Csala and Sgouridis 2013).

In general, the simulation results of the model suggested that if one of the manufacturers decides to innovate, then other manufacturers would follow with the delay thereby providing the first mover advantage to the manufacturer. The author commented on their results by pointing out the difficulty in the transition of the industry as radical innovation would be beneficial for both innovators, the competitive market and the industry as a whole. However, currently, the industry does not seem to be in harmony with these results as innovation in the commercial air transport industry is a very slow and incremental process (Csala and Sgouridis 2013). Nevertheless, this study provided a deep inside into the competitive strategies and the external effects of fuel prices on the future of the system using the system dynamics model.

(Abrantes et al. 2021) used a fleet system dynamics model (FSDM) to quantify the future reduction of CO_2 emissions by 2050 in the aviation system. The authors included imminent aircraft technologies and sustainable aviation fuel as a means to achieve low emissions in the future ATS.

The authors conducted seven scenarios for the two types of technology adoption in the system. Their results suggested that the IATA net zero goals cannot be achieved by only a single type of technology. On the contrary, the authors suggested that the combination of imminent aircraft technologies and the use of SAF together will be required to achieve higher emission reduction goals. Specifically, the SAF would need to increase 15% annually between 2030 and 2050 to help achieve the net zero goals. Furthermore, the imminent aircraft, technologies can only reduce up to 15% in CO_2 emissions compared to the base case scenario.

(Abrantes et al. 2024) conducted an analysis by considering the effect of the introduction of disruptive technologies concepts on aviation CO_2 emissions. The authors used a system dynamics approach to test the effects of technological progress on the future air transport system. Their model consisted of aircraft manufacturing production capabilities, current projections and forecasts. The authors included different types of technologies in the model and their specific entry into service targets. The different types of technologies that were considered for their study are showcased in Table 3.2.

Table 3.2 Overview of technologies considered by Abrantes et al. 2024 in their study

Group	Concept
New aircraft concept	Blended wing body
	Small BWB
	Strut braced wing
	Double bubble fuselage
	Strut braced wing with open rotor
	Boundary layer ingestion
Propulsion technology	Advanced turbofan
	Open rotor
	Ultrafan
Electric aircraft	Zunum aero 50
	Zunum aero 100
	ATR-72 fully electric
	Embraer 180 fully electric
Aerodynamics technology	Natural laminar flow

In the analysis, the authors considered five scenarios including business as usual scenarios. In the scenarios, different types of concepts were considered for introduction into the model according to their emissions reduction potential. The results of the analysis suggested that aircraft production capacity and the entry into service year are the two most important variables. Specifically, a 15% increase in the production capacity could lead to between 1 and 2.6% emissions reduction depending on the case and scenario (Abrantes et al. 2024). According to the authors, electric aircraft cannot contribute largely to emissions reduction because the electric aircraft considered in their analysis was for the regional flight segment where the amount of flights is less than other flight segments.

3.3 System Dynamics and Policy Analysis

The usage of system dynamics has been quite common in assessing various policy scenarios and their implication on the variables of interest. For instance, (Rozman et al. 2013) developed a system dynamics model to analyze organic farm development in order to support the policy decision making process in Slovenia. The model was used to simulate different policy scenarios for organic farming and their impact on the economic and environmental parameters of organic production at an aggregated level. With this study, the authors were able to provide policy-makers much needed clarity in planning strategies for the sustainable development of organic farming. (Espinoza et al. 2017) demonstrated the possibility of using system dynamics models for examining the policy support instruments for creating and managing the biodiesel market in Columbia. (Samara et al. 2012) developed a national innovation system (NIS) model to scrutinize the various aspects of innovation. With this model, the authors analyze the impact of various innovation policies on the NIS performance. The authors argued that their SD model was used as an experimental tool to conduct comprehensive simulation scenarios concerning innovation policies (Samara et al. 2012). (Kapmeier and Gonçalves 2018) assess the relationship between tourism-driven economic growth and environmental degradation with the help of a system dynamics model. The purpose was to use the model to suggest implications of various policies that could address the waste management issue resulting from tourism activities. (Jin et al. 2009) used system dynamics to develop an ecological footprint (EF) forecasting framework. The authors used this framework to support policy making for urban sustainability improvement by testing four policy scenarios.

The above suggests that system dynamics can be effectively used to analyze the impact of policy scenarios on the system of interest (Sterman 2000). This is because the inclusion of policy scenarios in the system dynamics model allows decision-makers to observe changes that can be entailed in the system (Ghaffarzadegan et al. 2011). Therefore, using a system dynamics approach to assess the impact of different policy scenarios to help achieve sustainability initiatives set by the institutions as discussed in Sect. 2.5.2 can provide decision-makers a differentiated understanding of policy effects on the overall ATS. Furthermore, using the system dynamics approach also brings the ability to model different policy scenarios by changing the key parameters and monitoring them easily to choose the best strategy to achieve the intended goal (Sterman 2000).

3.4 Research in the Air Transport System

Studies that have analyzed the future of the ATS from a high level viewpoint in the presence of new technologies are scarce in the literature. For instance, (Eurocontrol 2022a) conducted a study to analyze the air traffic development by 2050 in the presence of different technologies and policy measures to reach net zero goals within the EU. The analysis was conducted by first, forecasting the different growth rate scenarios of number of flights and second, taking into account future airport capacity constraints and lastly, modeling fuel burn to examine the CO_2 emissions. Their results suggest that to achieve net zero emissions in aviation in the EU by 2050, 279 million tons of CO_2 would need to be abated. Furthermore, for the net zero goal to be achieved, 17% emissions reduction would need to come from more efficient conventional aircraft, 2% from electric or hydrogen powered aircraft, 8% from efficient air traffic management (ATM) and airline operations, 41% from sustainable aviation fuels (SAFs) and lastly 32% from other measures. The main drawback of this analysis is that the study only considered different scenarios of what could be the different share of aspects to reduce emissions but does not discuss any requirements for the same. Furthermore, the base of the different settings of the scenarios is not clearly discussed. Another study from Eurocontrol examines the role of SAFs in the "Fit for 55" plan initiated by the EU to reduce emissions by 55% by 2030 (Eurocontrol 2022b). The study is conducted by using Eurocontrol's existing database and generating different scenarios to observe results. The main findings of this report are that a 55% CO_2 emissions reduction target by 2030 is achievable in all scenarios, but this relies heavily on market based measures. Specifically, the high scenario with the most traffic is counterintuitively the most efficient to reach net zero emissions by 2050 at lower cost, as higher revenues will drive increased investment in new technology.

(O'Connell et al. 2019) conducted an analysis to consider GHG emission reduction concerning energy balances using biofuels in aviation. The article's main aim was to consider whether the emission reduction benefits are justified in terms of the energy efficiency of each alternative. The results were then aimed at decision-makers on the informative investment options in this sector. Furthermore, the study provided insights into the role of biofuels in recent legislation related to biofuels in aviation to reduce emissions reduction. The result of their analysis suggested that the second-generation biofuels are promising in terms of their greenhouse gas emission reduction but at the same time their overall energy efficiency performance is relatively poor.

(Kousoulidou and Lonza 2016) conducted a scenario-based analysis to analyze the future growth in the European aviation sector. The authors assess the future growth in terms of fuel demand and the role of biofuels in fulfilling the same. Most of their scenarios were aimed at finding CO_2 emissions until 2030. The authors overall conducted three scenarios based on different parameter values and traffic volumes. The results of the scenarios suggested that the emission outcomes can be 360 million tons, 271 million tons and 401 million tons of CO_2 emissions from aviation in central pessimistic and optimistic scenarios, respectively. Furthermore, in their analysis regarding different types of biofuels, their results suggested that most of the demands until 2030 will be served by HEFA production type in the EU, however after 2030, the demand could be fulfilled by using FT-synthesis.

(Åkerman 2005) used a backcasting method to analyze different future scenarios for emission reduction for air transport until 2050 using new types of aircraft concepts. The author suggested that the conventional turbofan aircraft will not be sufficient to reach even the less demanding climate emissions reduction goals even if a 40% reduction of fuel intensity is achieved. On the other hand, developing more radical aircraft configurations, for example, the flying wing is believed to have the greatest potential for reducing fuel consumption and thereby emissions. According to the author, the least risky approach would be to develop a high-speed propeller aircraft with a lower cruise speed. According to the author's analysis, this type of aircraft would provide a 56% reduction in carbon emissions per passenger kilometer compared to 2000-year value. But on an overall level, even with the introduction of this technology, weakened economic growth would be required to reach the most demanding climate goals for the aviation system.

From a qualitative point of view, (Ghosh et al. 2017) conducted a study presenting a quantitative scenario philosophy for systems analysis and innovative concept design within the air transportation system (ATS) context. The authors used an "atomic model" consisting of main stakeholders such as manufacturers, airlines, air navigation service providers (ANSPs) and airports. Moreover, they employ a waterfall model that illustrates the mental model of integration and decomposition across cascades of levels of detail, ranging from the global scenario level to individual technologies. The above mentioned studies are either qualitative or scenario based analysis for the future ATS. One of the drawbacks of these studies is that they lack the assessment of the effects of manufacturing delays, fuel production delays, interlinkages between stakeholders and overall requirements of fuel and electricity to achieve goals.

Air Transport System Model 4

In this section, the modeling methodology and extensions of the air transport system (ATS) model required for this thesis are explained. The modeling section is divided into three parts, in the first part the goals of the extended model are explained. This part is particularly important in understanding the reasons for extending the current ATS model. In the second part, the model extensions are discussed in more detail. This includes explaining aspects related to system dynamics modeling such as developing causal loop diagrams of the mental models, stock and flow diagrams, and formulation of the model in the form of equations. Lastly, in the third part, the model validation and testing are discussed. In this part, the model is tested based on best system dynamics methodology practices to validate the model.

4.1 Air Transport System Model Goals

The previous ATS model discussed in Chap. 3 consisted of the sub-sections air travel demand, airline industry, aircraft manufacturers and fuel producers. However, the previous models did not include aspects like new technologies, new types of fuels, mechanisms of different policies and different range segments. For this reason, the previous model can be used as a base but needs to be extended to answer questions for this thesis. Specifically, to incorporate the new propulsion concepts like battery electric, hydrogen combustion and hydrogen electric aircraft in the simulation model, the fleet planning and operation mechanism in the airline

C. Talwar, *Transition Pathways towards a Low Emission Air Transport System of the Future – A System Level Analysis*, Produktion und Logistik,
https://doi.org/10.1007/978-3-658-46507-0_4

industry sub-section would need to be extended. The aircraft manufacturing sub-section would need to be extended with capacities related to the new propulsion concepts. Also, since the airline industry heavily depends on costs to operate the aircraft, therefore, the costs related to new propulsion concepts would also need to be added. Regarding the new type of aircraft fuels such as SAF and hydrogen, extensions related to fuel production capacities would need to be done. To test for various policies that influence the adoption rate of new propulsion concepts, a new policy mechanism would need to be added to the ATS model. Lastly, the previous model would need to be extended in terms of flight segments by adding subscripts of different range segments.

The first goal of the extended model is to test for different adoption strategies of new technologies and fuels. Testing different strategies in short, medium and long-haul flight segments would help in assessing their effects on the system. Understanding the different effects would provide clarity to the policy-makers, decision-makers and stakeholders about the potential of the new technologies in different range segments. The second goal of the extended model is to assess the different policies that can influence the adoption rate of new propulsion technologies. This would provide clarity on the extent to which in-flight emissions can be reduced with the help of policies and an increase in the adoption of new technologies. Specifically, it would help in assessing which type of emissions (such as CO_2) the new technologies can help to reduce from the environment under different policy settings. The third goal of the extended model is to showcase the different requirements of the system with respect to the strategies. Specifically, the model would be able to showcase the amount of SAF, hydrogen and green electricity that would be required to fulfill the needs of the system if certain strategies are chosen by the decision-makers. This would particularly provide clarity on the uncertainty of the requirements for the system to reach emissions reduction goals.

4.2 Air Transport System Model Extensions

In this section, the extensions done for the existing ATS to meet the goals of this thesis are discussed in detail. In the first part, the basic structure of the extended ATS model is presented. The basic structure provides a conceptual model that helps in understanding the boundary and new components that are included in the model. In the second part, the explanation of the mental models in the form of causal loops and stock and flow diagrams is done. Lastly, the third part consists of the underlying equations of the extended model.

4.2.1 Basic Structure of the Extended ATS Model

The system model is based on the work of (Kieckhäfer et al., 2018) and (Pierson and Sterman 2013). However, as discussed in Chap. 3, the previously developed model is not enough to meet the goals of this thesis; therefore, the extensions in the model regarding the introduction of battery electric, hydrogen electric and hydrogen combustion aircraft, different policy mechanisms, range segments and hydrogen and SAF fuel production along with green electricity requirements are done. Figure 4.1 showcases an overview of the extended model in the form of a conceptual model. In the figure, the brown colored text represents the extensions done in the existing model. The scope and the boundary of the model is defined by the outermost dashed line. Arrows pointing inward depict exogenous parameters. The interaction between air travel demand, airline industry and aircraft manufacturers generates information about the system given by arrows pointing outwards.

The system's functioning starts with the generation of air travel demand, influenced by various factors such as the impact of ticket prices, GDP per capita, population size, and congestion levels. The air travel demand is generated with the help of both endogenous and exogenous variables. The air travel demand is then processed by the airlines to calculate the required fleet. Based on these calculations, the airlines constantly adapt their available seat kilometers (ASK) to demand developments by changing fleet size and fares. To expand their fleet, the airline orders new aircraft from the manufacturer. Depending on the cost advantage, policy implication and possible applications, the airlines order conventional, electric, hydrogen combustion or hydrogen electric aircraft. After taking into account production capacities and delivery bottlenecks, orders placed are realized by the aircraft manufacturers. The delivered aircraft are then operated by the airlines. The airlines then choose the type of aircraft for the range segment depending on the demand in each segment. After receiving the fleet ordering information, the fuel producers adjust their capacities according to the requirements. There are now two different types of fuel producers introduced, for sustainable aviation fuel and hydrogen as jet fuel. The operating margin in the system is generated with the help of revenue and cost of operation of the fleet obtained together. The airline fares are influenced by the effect of competition, load factor and operating costs. The air travel demand is in turn affected by the airline fares along with the congestion. In this way, the system forms a feedback mechanism that serves as a base for its cyclical dynamics (Liehr et al. 2001).

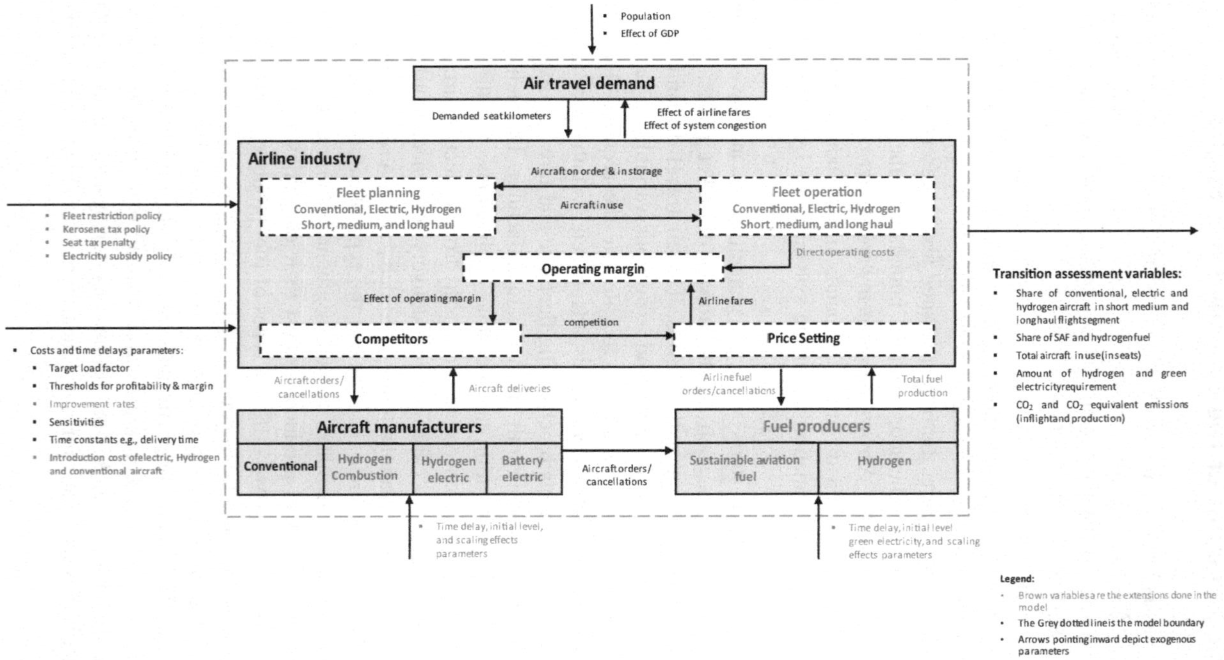

Fig. 4.1 Overview of the extended air transport system model

In the above mentioned process, the CO_2 and CO_2 equivalent emissions are generated based on the active conventional, hydrogen and hydrogen electric fleet. Also, the CO_2 and CO_2 equivalent emissions from fuel production of hydrogen and SAF are taken into account. Policy measures like taxes, subsidies and a fleet restriction are integrated into the system. The kerosene tax directly increases the costs to introduce and operate conventional aircraft, whereas electricity subsidy reduces the costs for electric aircraft. A fleet restriction policy directly affects the fleet planning mechanism of airlines, by forcing them to have a certain percentage of their fleet be electric. In this way, the considered structure of the system functions in dynamic interconnected feedback loops.

4.2.2 Causal Loop and Stock and Flow Diagrams

In order to understand the extensions made in the model presented in Sect. 4.2.1, the mental models in the form of causal loops and stock and flow diagrams need to be explained. In a dynamic simulation model, interactions are of high importance, therefore, for the ATS the feedback structure between the variables is explained with the help of causal loop diagrams. The causal loop diagram (CLD) is a tool that helps in showcasing the causal feedback loop structure between the different variables of interest in a system (Sterman 2000). The different variables from the sub-system interact with each other and affect the adoption rates of new technologies. Representing the system in this way facilitates a better understanding of the initial impression of possibilities with respect to the feedback structure within the system. Although there are many aspects related to the extensions done in the model that can be shown in the form of causal loops and stock and flow diagrams, only the most important ones are discussed. All the diagrams related to extensions are shown in the form of a figure, Table 4.1 shows the meaning of all symbols for all the figures in this section. The causal loop and stock and flow diagrams for the different sub-sections of the system are explained in the following sub-sections. In the diagrams, the variables for hydrogen denote both hydrogen based propulsion technology, i.e., hydrogen electric and hydrogen combustion technology, where hydrogen electric is aimed at short-haul flight segment and hydrogen combustion at medium and long-haul flight segments.

Table 4.1 Explanation of symbols for causal loop and stock and flow diagrams

Symbol	Meaning
+	Positive/Same direction movement/directly proportional
–	Negative/Opposite direction movement/Indirectly proportional
II	Delay in transmission of information or material
Bolded variables	Exogenous inputs based on data
Green variables	Used more than once in diagrams to connect across diagrams

4.2.2.1 Airline Industry

In the airline industry, one of the most important aspects is the fleet operation mechanism. The fleet operation mechanism consists of feedback loops that represent the process of airlines for ordering, operating and retiring the aircraft fleet. The extension made in the model after the introduction of battery electric, hydrogen based propulsion aircraft is explained first in the form of causal loop diagrams and then secondly in their corresponding stock and flow diagrams.

Fleet operation overview: causal loop diagram (CLD)

Figure 4.2 and Figure 4.3 showcase the extensions done in the model for the introduction of hydrogen and electric aircraft with the help of causal loop diagrams (CLDs). The diagram consists of different balancing and reinforcing feedback loops for the fleet operation segment. In Fig. 4.2 and Fig. 4.3, the variables in black are the preexisting variables from the previous model feedback structures. The new extensions done in the existing feedback structure of the model are underlined and italicized. The bolded variables are the exogenous parameter introduced to extend the model. Finally, the variables marked in green are the variables that are used more than once in the causal loop diagrams to interlink the sub-systems.

In both diagrams, the causal loop function starts with an increase in air travel demand which increases the total desired capacity orders. The desired capacity represents the action taken by airlines to increase their aircraft orders to match the demand. The increase in demand is balanced in both cases by balancing loop *B1*. Figure 4.2 shows the extensions in the model regarding the hydrogen fleet. The extensions are implemented by dividing the desired capacity into different types of fleet with hydrogen propulsion based fleet being one part and conventional the other. The extent to which the share of the fleet is chosen depends on the hydrogen share multiplier which is the potential percentage share of hydrogen aircraft in the system by 2050 based on expert opinions and industry reports (Fuel Cells and Hydrogen Joint Undertaking 2020; Hoelzen et al. 2022a). Furthermore, the

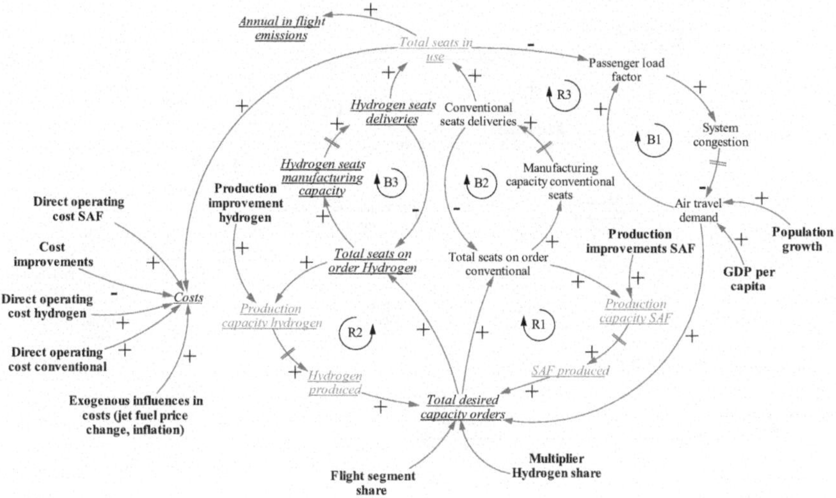

Fig. 4.2 Causal loop diagram of the extended fleet operation segment for hydrogen aircraft and SAF

extension in the model regarding flight segment is done at this point by feeding flight share data (Hoelzen et al. 2022a) to the model. This is done in two steps, first, creating four different types of independent flight segments namely overall, short-haul, medium-haul and long-haul. Second, after creating the four independent segments each segment is fed with the respective share in the system according to the exogenous input data. In Fig. 4.2, the reinforcing loop generated by the increase in hydrogen fuel production capacity as the airlines choose to order more hydrogen based aircraft is represented by *R2*. The rationale for this loop comes from the fact that an increase in orders by the airlines causes the fuel producers to have the incentive to produce more as the expectation of usage of hydrogen fuel based aircraft in the future increases. The reinforcing effect of loop *R2* is balanced by balancing loop *B3*, this balancing loop helps to not have unrealistic exponential hydrogen fuel production growth. In loop *B3*, the increase in total seats on order hydrogen increases the hydrogen seats manufacturing capacity and the amount of hydrogen seats deliveries. The increase in deliveries negatively affects the total seats on order hydrogen which represents the adjustments done by the airlines to not over-order the fleet. Likewise, in the case of Sustainable Aviation Fuel (SAF), any rise in the reinforcing loop *R1* is counteracted by the adjustments made by airlines within the

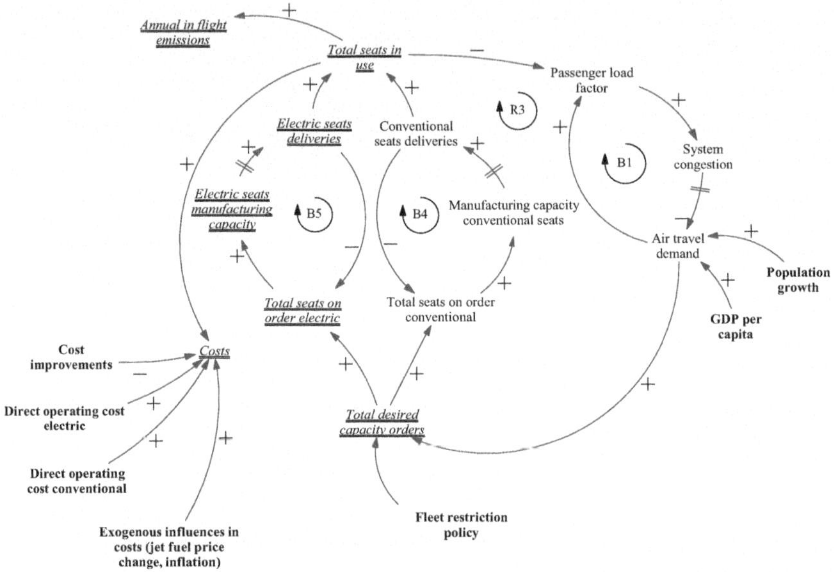

Fig. 4.3 Causal loop diagram of the extended fleet operation segment for electric aircraft

balancing loop *B2*. In the case of electric aircraft, the reinforcing loop is replaced by the fleet restriction policy. In the case of electric aircraft as seen in Fig. 4.3, when the fleet restriction policy is active then the airlines are forced to operate a certain percentage of their fleet with battery electric aircraft which increases the electric aircraft adoption rates. The increase in adoption can be compared to the reinforcing loop *R2*. Similar to balancing loops *B2* and *B3*, balancing loops *B4* and *B5* balance the increase in the fleet as a result of either an increase in demand or fleet restriction policy. Lastly, the reinforcing loop *R3* shown in both diagrams shows the gradual increase in the fleet size ordered by the airlines as the demand rises, in other words, showing the growth of the industry over time.

Fleet operation overview: stock and flow diagram

Figure 4.4 and Figure 4.5 represent the mechanism for electric and hydrogen fleet operation in the form of stock and flow diagrams. For simplicity, the mechanism for both technologies is kept the same. The indicated order rate electric represents the amount of new fleet needed to be ordered for electric aircraft. On the other hand, the indicated order rate hydrogen represents the amount of new hydrogen

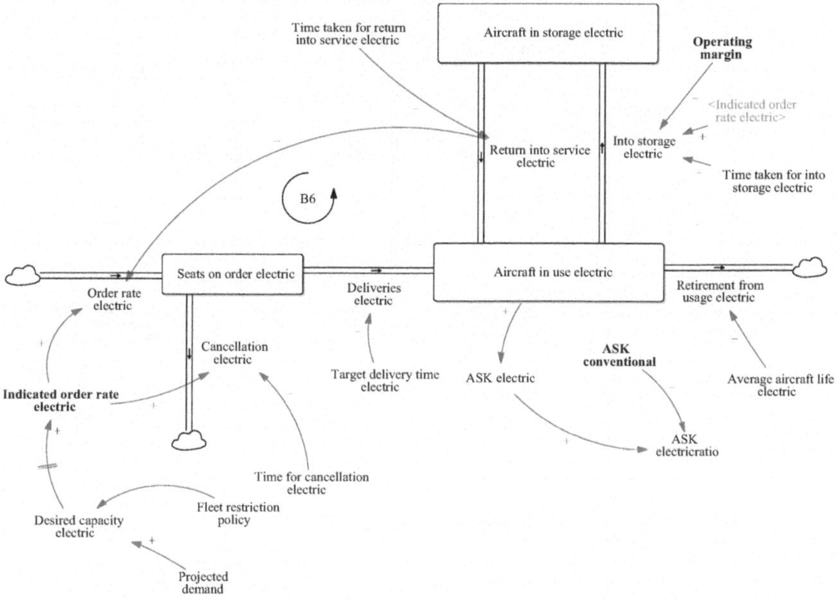

Fig. 4.4 Stock and flow diagram fleet operation for electric aircraft

fleet that needs to be ordered to meet the demand. The order rate before reaching the manufacturers is adjusted for the seats that will be entering into service from storage soon. Over here, the order is considered a rate that flows into the stock seats on order thereby filling it over time. Furthermore, the figure also shows the mechanism for the cancelation of the fleet. The outflowing rate of cancelations depends on the indicated order rate. A negative indicated order rate denotes that the airlines already have too much fleet and need to cancel the ones that have been already ordered. This particular mechanism works as a double check for the airlines to make sure the optimum amount of fleet exists in usage. Once the orders reach the manufacturer then, they start to process the orders based on their capacity which is explained in Sect. 4.2.2.2 below. When the aircraft are delivered from the manufacturers, the airlines make the decisions regarding the fleet operations. The aircraft which are in operation are named under the stock aircraft in use. The stock of aircraft in use is utilized to obtain the variable available seat kilometer (ASK) by multiplying it by the number of kilometers each seat travels in a given year. If there is a situation where airlines make losses, then they downgrade their fleet by initiating storage as

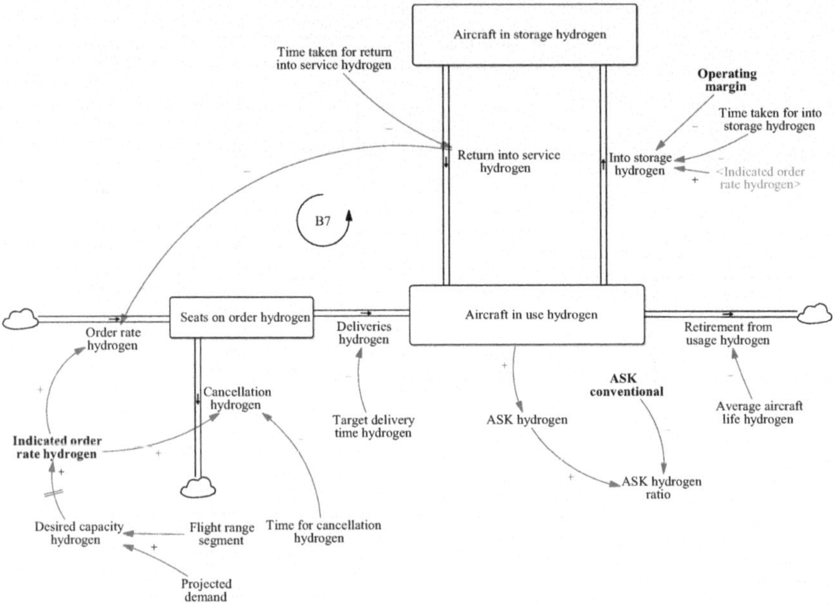

Fig. 4.5 Stock and flow diagram fleet operation for hydrogen aircraft

a safeguard against the excess losses. These aircraft come back into service if the airlines are able to again make profits and need to increase their fleet operations. In the process of fleet operation, there is one balancing loop each for electric and hydrogen propulsion concepts that are formed. The balancing loops *B6* and *B7* indicate that an increase in order rate would increase the aircraft in use. A further increase in aircraft from storage coming back into service would then again reduce the order rate itself. This kind of mechanism would serve as a balancing loop so that the seats are not over ordered.

4.2.2.2 Aircraft Manufacturers

Figure 4.6 and Figure 4.7 explain the mechanism for the electric and hydrogen aircraft manufacturers in the extended model. To deliver the orders, the aircraft manufacturers set a target for their manufacturing capacity based on the seats on order and the expected time taken for delivery of these orders. As can be seen in Fig. 4.6 and Fig. 4.7, the target is adjusted with the actual manufacturing capacity that the manufacturers have at a given time. Thus, the variables manufacturing capacity adjustment, manufacturing capacity change along with the stock aircraft manufacturing capacity form a balancing feedback loops *B8* and *B9*.

This balancing feedback loop actively adjusts to reach the manufacturing capacities required to fulfill the orders. For both new technologies, the manufacturing capacity is added with an improvement rate. This is done because the start of the manufacturing process of the new type of aircraft cannot be increased quickly. Thus, the improvement rate accounts for improvement in manufacturing know-how which reduces the time taken to adjust for manufacturing capacity, thereby making the process more efficient over time. This particular aspect is added to the model with an assumption that after the introduction of electric and hydrogen based aircraft the manufacturers will work towards creating higher efficiencies and productivity of production cycles.

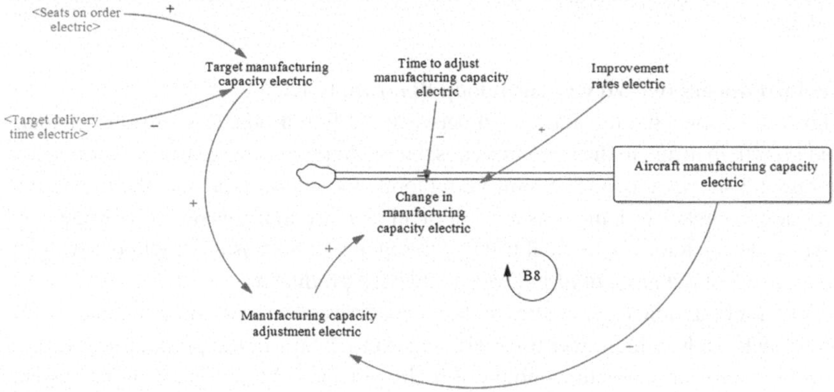

Fig. 4.6 Stock and flow diagram for electric aircraft manufacturers

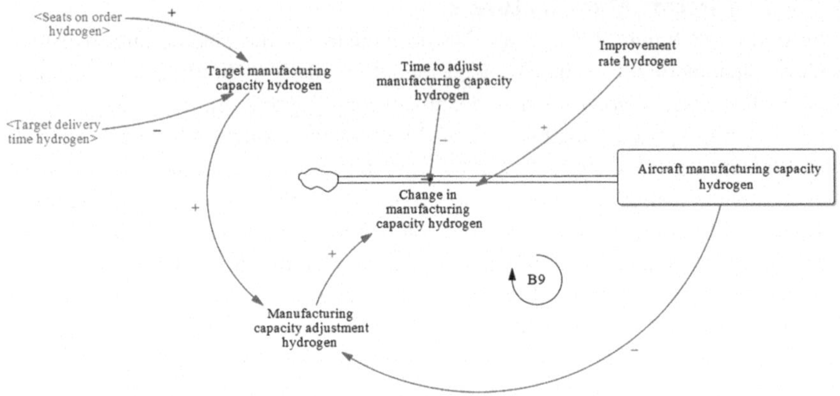

Fig. 4.7 Stock and flow diagram for hydrogen aircraft manufacturers

4.2.2.3 Aircraft Fuel Producers

The fuel producers sub-section is connected to the airline industry sub-section because an increase in the adoption of either hydrogen or SAF would increase the incentive for fuel producers to increase their capacities. The depiction of mental models in the form of causal loops and stock and flow diagrams are shown as follows:

Fuel producers overview: causal loop diagram (CLD)

Figure 4.8 shows the feedback loop between the fuel producer's capacity changes and growth in usage in the form of a causal loop diagram. This diagram is connected to Fig. 4.2 shown above. The reinforcing loops *R4* and *R5* represent the increase in production capacity changes as a result of an increase in the usage of hydrogen and SAF in the system. The annual use of hydrogen and SAF is affected positively by an increase in the amount of hydrogen and SAF produced.

The fuel manufacturers increase their capacities as a result of an increase in the use of new fuels, which over time with regard to delays in the production capacity developments increases the amount of hydrogen and SAF produced in the system. The reinforcing feedback loops *R4* and *R5* are counteracted by the balancing feedback loops *B10* and *B11*, depicting the decrease in the growth of hydrogen and SAF fuel demands as these technologies are increasingly adopted over time. The reinforcing loop *R4* and balancing loop *B10* together form a goal seeking behavior in the system (Sterman 2000). In a goal seeking behavior, the variable or product of interest reaches its growth due to a reinforcing loop and constant correction through

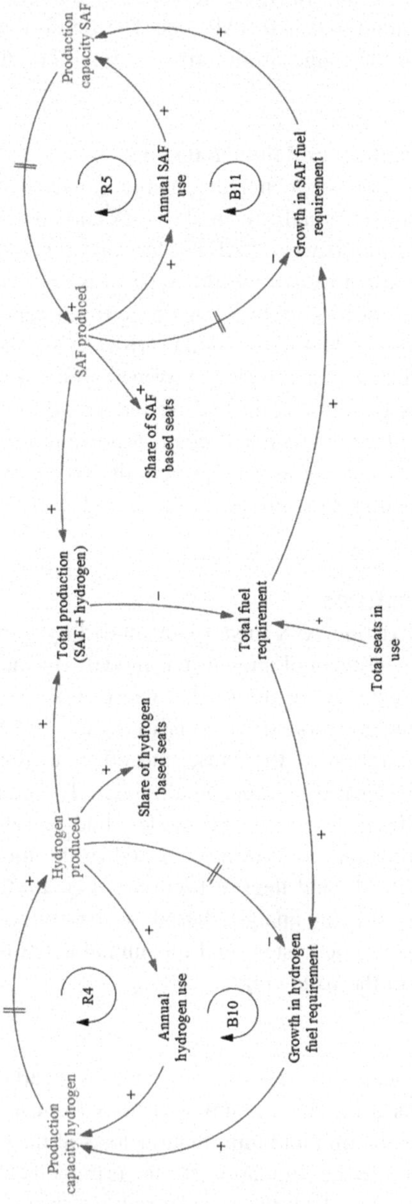

Fig. 4.8 Causal loop diagram showcasing feedback loop between fuel producers and capacity growth

balancing feedback loops. In this study, the variables of interest are the adoption of hydrogen fuel, hydrogen aircraft and SAF fuel. The reinforcing loop for these variables is the increase in the usage of aircraft over time, i.e., the growth of the industry.

Fuel producers overview: stock and flow diagram

Figure 4.9 shows the connection between hydrogen fuel capacity and green electricity availability for green hydrogen in the form of a stock and flow diagram, whereas Fig. 4.10 shows the connection between SAF production capacity, green hydrogen capacity and electricity availability for the PtL type of SAF. The balancing loop that represents the adjustment in growth for hydrogen fuel capacity is shown in Fig. 4.9 labeled as *B12* with a clockwise circular arrow. Also, the balancing loop for the adjustment of the change in green electricity availability is showcased by the balancing loop *B13*. Similarly, for SAF, the hydrogen required for the PtL type SAF and the electricity required for the green hydrogen is shown in Fig. 4.10, labeled as *B14*, *B15* and *B16*. These balancing loops represent the goal seeking behavior for hydrogen fuel, SAF, hydrogen for the PtL type of SAF and electricity requirements.

4.2.2.4 Policy Implications

In terms of policies, the existing ATS model is extended by introducing policies after the introduction of new technologies in the model. By introducing various policy scenarios, certain segments in the model, i.e., cost structures, order rates, fleet operation, and annual aviation CO_2 emissions, are affected. Figure 4.11 shows a comprehensive diagram of the cause and effect in the system when a policy is introduced in the form of a conceptual model. The aim of this conceptual model is to project the different mental models that represent the potential behavior of the system after the policies are enacted in the model. The jet fuel tax, seat tax, electricity subsidy and fleet restriction policy are the different policies that have been chosen for this analysis based on the review in Chap. 2. The effect that each policy has on the system and the annual aviation CO_2 emissions is explained individually in the following.

Jet fuel tax

Jet fuel tax is a policy in which the government imposes a certain amount of tax on airline jet fuel. This is done to discourage airlines from using conventional aircraft. The policy affects the unit and introduction costs of the conventional fleet by increasing the fuel prices. In the simulation, the tax is implemented starting in the year 2025. As can be seen in Fig. 4.11, the levied tax affects the costs positively, i.e.,

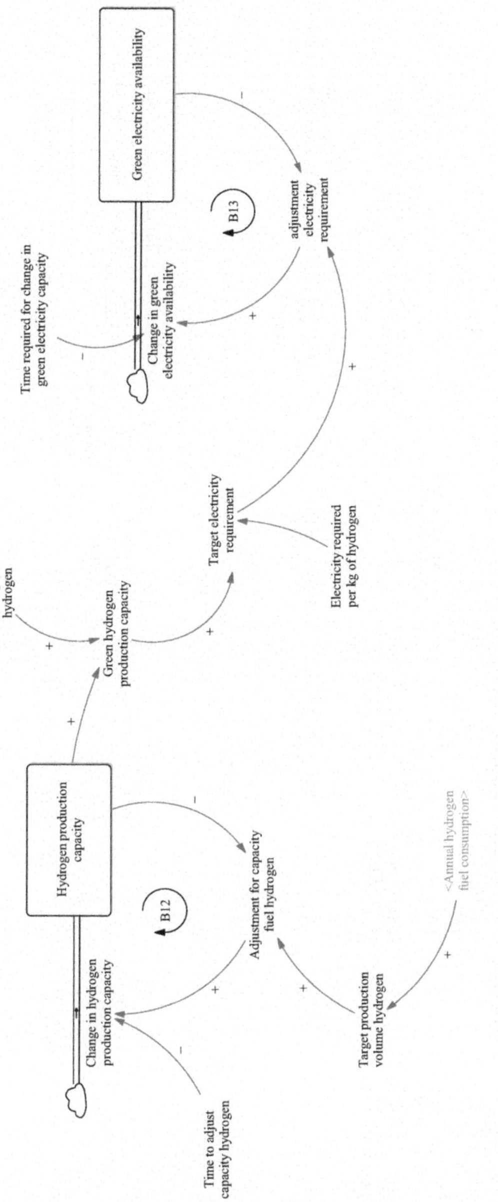

Fig. 4.9 Stock and flow diagram for hydrogen fuel production capacity

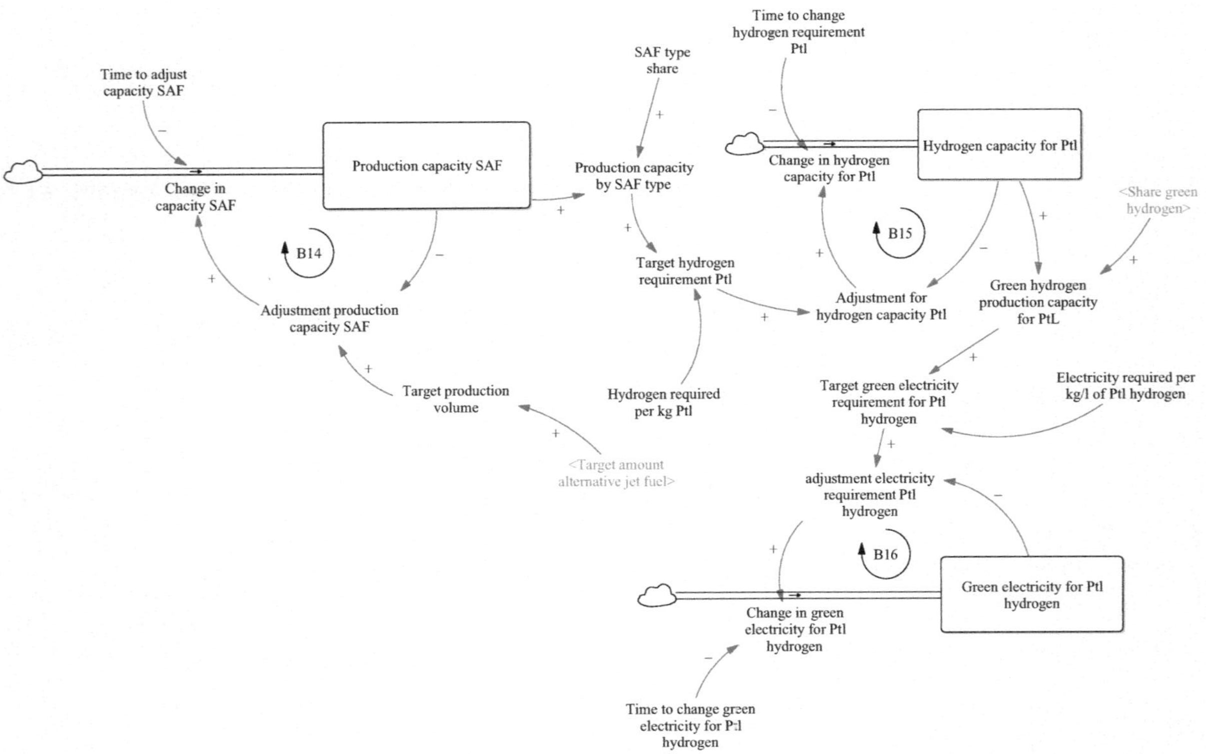

Fig. 4.10 Stock and flow diagram for SAF fuel production capacity

the increase in costs due to the tax implementation will discourage the airlines from ordering more conventional fleets. The effect of this in turn reduces the conventional aircraft in use and available seat kilometers for the conventional fleet, which would mean a reduction in fuel usage and ultimately reduces the annual aviation CO_2 emissions.

Electricity subsidy

For electricity subsidy, the government provides a certain amount of subsidy on the electricity cost for the battery electric fleet[1]. The subvention reduces the electricity costs for charging the planes, thereby promoting the new technology.

The electricity subsidy is introduced in the year 2035 in the model. In Fig. 4.11, the electricity subvention policy negatively affects the unit and introduction costs for the electric fleet, thus encouraging the airlines to order more electric aircraft. The increase in orders would increase electric aircraft usage and thereby the available seat kilometers for the electric fleet. An increase in electric fleet reduces the usage of conventional fleet, thereby lowering the annual aviation CO_2 emissions.

Seat tax

In this policy scenario, the government imposes a certain amount of tax on every seat that is operated by the airlines in a conventional aircraft. The seat tax increases the unit and introduction costs for conventional aircraft, thereby discouraging the usage of the same. The seat tax policy is introduced in the year 2038 in the model. As can be seen in Fig. 4.11, the implementation of seat tax increases the comparative cost of operating conventional aircraft. The increase in costs decreases the conventional fleet orders. The decrease in future orders ultimately reduces the available seat kilometers for conventional fleet. The above in turn facilitates the reduction of the annual aviation CO_2 emissions.

Government fleet restriction policy

The fleet restriction policy refers to a scenario where the government specifies a certain percentage of the total fleet that needs to be operated by battery electric aircraft obligatorily. In the model, the government fleet policy is introduced in the year 2038. In this scenario, the first step that the airlines take after the policy has been introduced is that they replace the conventional fleet with an electric one. The airlines implement this by placing a portion of their operational conventional fleet

[1] Considering the recent measures taken by the EU commission, the electricity subsidy policy is tested in the model with an assumption that the energy crisis in the EU will be resolved by 2035—https://commission.europa.eu/strategy-and-policy/priorities-2019-2024/european-green-deal/eu-action-address-energy-crisis_en.

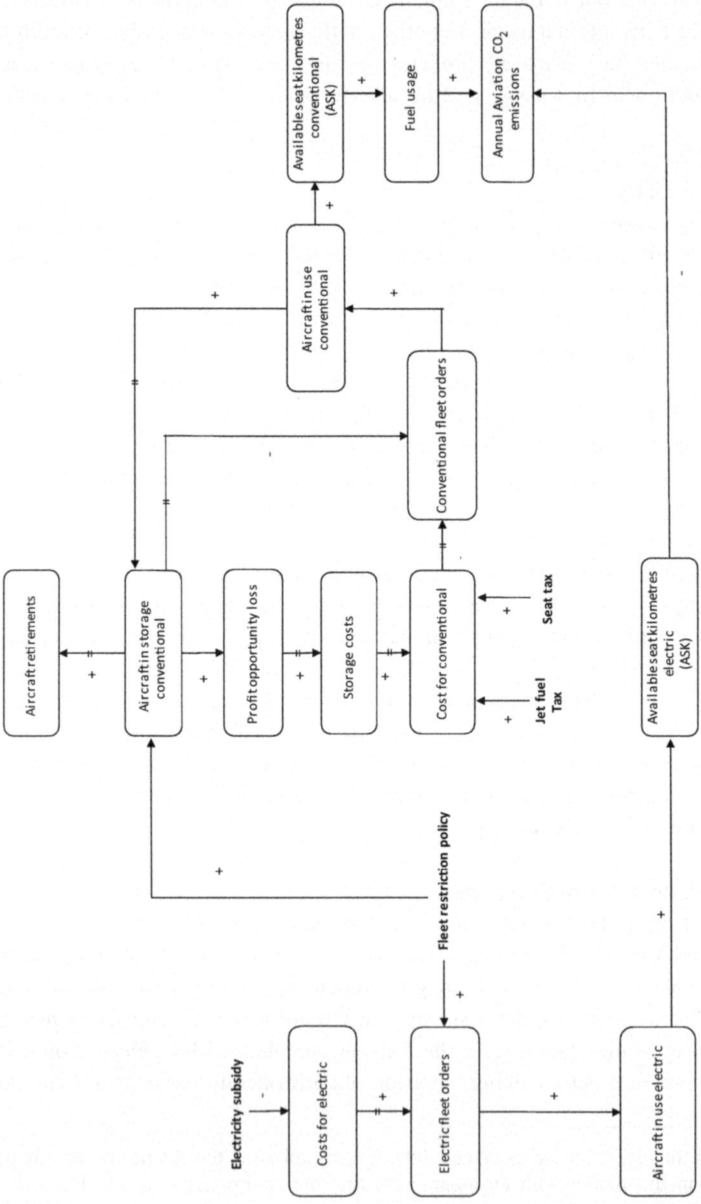

Fig. 4.11 Implication of various policy scenarios in the ATS model

into storage. At the same time, the airlines order the same amount of electric seats to compensate for the reduction in the operational fleet. This particular process is not immediate but takes place with delays for ordering and manufacturing electric aircraft and for aircraft going into storage. As can be seen in Fig. 4.11, when the conventional aircraft starts to accumulate in storage, it has a three-fold effect on the system. First, the increase in the profit opportunity losses increases the storage costs for the airlines. This forces the airlines to reduce their orders for conventional future aircraft. The reduction in conventional orders reduces the conventional aircraft in use and ultimately the annual aviation CO_2 emissions. Second, with the increase in the number of aircraft in storage, the airlines will decide to retire these aircraft after a certain point. They make this decision because the aircraft in storage incur high costs. Third, the implementation of a fleet restriction policy increases the electric fleet order rate. As the airlines are forced to shift toward more electric aircraft, the increase in orders ultimately increases the operation of the electric fleet.

4.2.3 Model Formulation and Equations

In order to understand the relationship between the variables and the effect of feedback loops, the underlying modeling equations need to be discussed. As the system dynamics methodology deals with changes in variables over time in the form of a simulation model, therefore, ordinary differential equations (ODEs) are often used to represent this mathematically. The main reason for this is that the differential equations are able to capture the rate of change in stocks correctly (Sterman 2000). In this section, the equations pertaining to the diagrams shown in Sect. 4.2.2 are explained. Explaining the equations this way provides more clarity on the way in which extensions are done in the model. Certain equations regarding the extensions for different sub-sections are explained in the technical report (Talwar et al. 2021b).

4.2.3.1 Airline Industry

As explained in the previous section, the most important aspect of the airline industry component of the model is the process of ordering and operation of new aircraft. Only the most important equations regarding the model are discussed here, more details regarding the extensions can be seen in the technical report (Talwar et al. 2021b).

New aircraft order rate

The modeling extensions both for hydrogen and electric aircraft start from the process of the new aircraft order rate mechanism. Over here, the airlines after assessing the demand calculate the amount of seats required. Due to the focus of this thesis and better explain the extensions, the steps regarding the demand forecast are not discussed over here but can be found in the technical report (Talwar et al. 2021b). The total desired capacity in seats is calculated by dividing the estimated demand by the target load factor for the airlines,

$$DC(t) = \frac{ED(t)}{TLF(t)}, \tag{4.1}$$

where $ED(t)$ is the estimated demand in seats and $TLF(t)$ is the target load factor. The separation of three segments in the model is done by

$$DC^i_{range\ split}(t) = DC(t) \cdot FSS^i, \tag{4.2}$$

where FSS^i is the input data that is fed into the model for all three ranges and the "i" represents each segment with $i = 1$ overall fleet, $i = 2$ short range, $i = 3$ medium range and $i = 4$ long range. The amount of new seats required for a particular technology is represented by the desired capacity, where $DC_e(t)$ is for electric and $DC^i_{hydrogen}(t)$ is for hydrogen.

The introduction of hydrogen based propulsion technology in the model is done by splitting the desired capacity for hydrogen and conventional aircraft, where the equation for hydrogen is

$$DC^i_{hydrogen}(t) = \begin{cases} 0, & t < t^i_{hydrogen} \\ DC^i_{range\ split}(t) \cdot Multiplier^i_{hydrogen}, & t \geq t^i_{hydrogen} \end{cases}, \tag{4.3}$$

where $t^i_{hydrogen}$ is the time to introduce hydrogen aircraft in the system for short, medium and long-haul flight segments. The $Multiplier_{hydrogen}$ is the potential percentage share of hydrogen aircraft in the system.

Similarly, the desired capacity for electric aircraft $DC_e(t)$ is also split but since in this model, only short-haul aircraft are taken into account the split is only done because of policies and not range segments.

$$DC_e(t) = \begin{cases} 0, & t < t_{short} \\ DC(t), & t > t_{short} \text{ and } IC_{per}^{diff}(t) \leq 0 \\ DC_e^{policy}(t), & t > t_{short},\ IC_{per}^{diff}(t) > 0 \text{ and } t > t_{policy} \\ 0, & else \end{cases}, \tag{4.4}$$

In this equation, t_{short} is the year when electric aircraft is introduced in the model and t_{policy} is the time when the government fleet restriction policy is introduced in the model. The explanation for $DC_e^{policy}(t)$ desired capacity for electric when the government fleet policy is introduced is given in Equation (4.79) and (4.80). The IC_{per}^{diff} refers to the perceived introduction costs difference to the airlines between the electric and conventional fleets. The difference in costs is calculated by subtracting the introduction costs of conventional aircraft from the electric aircraft.

Thus, if $IC^{diff}(t) > 0$, then it implies that electric aircraft are more expensive than conventional ones. On the other hand, If $IC^{diff}(t) < 0$ then electric aircraft would be less expensive to introduce.

The reason for introducing perception delay is that because of the introduction of the electric aircraft, the airlines would require some time to perceive the introduction costs differences so as to adjust their future orders.

Together with the currently used aircraft $AU_{hydrogen}^{i}(t)$, the capacity adjustment $CA_{hydrogen}^{i}(t)$ for hydrogen aircraft is

$$CA_{hydrogen}^{i}(t) = \frac{DC_{hydrogen}^{i}(t) - AU_{hydrogen}^{i}(t)}{\tau^{CA}}. \tag{4.5}$$

The adjustment for the fleet capacity for the electric fleet is given by

$$CA_e(t) = \frac{DC_e^{per}(t) - AU_e(t)}{\tau^{CA}}. \tag{4.6}$$

Over here the only difference between the two equations is that the perceived desired capacity $DC_e^{per}(t)$ refers to the delay that the airline would require to adjust for the capacity requirements. Whereas, in the case of hydrogen this is not present because the policy adjustments are not done in that model.

Furthermore, the capacity growth adjustment for hydrogen $CA_{hydrogen}^{G,i}(\text{t})$ is calculated as

$$CA_{hydrogen}^{G,i}(\text{t}) = AU_{hydrogen}^{i}(t) \cdot g^{expected}(t) \cdot W^{G}, \tag{4.7}$$

where, W^G is the weighting factor, $g^{expected}(t)$ is the expected growth rate of demand and $AU^i_{hydrogen}(t)$ the current hydrogen aircraft in use.

Similarly, the capacity growth adjustment for electric aircraft is given as

$$CA^G_e(\text{t}) = AU_e(t) \cdot g^{expected}(t) \cdot W^G. \tag{4.8}$$

The desired acquisition rate for hydrogen $DAR^i_{hydrogen}(t)$ is calculated using the capacity adjustment growth $CA^{G,i}_{hydrogen}(\text{t})$ and capacity adjustment $CA^i_{hydrogen}(t)$

$$DAR^i_{hydrogen}(t) = CA^i_{hydrogen}(t) + CA^{G,i}_{hydrogen}(\text{t}). \tag{4.9}$$

The desired acquisition rate for electric $DAR_e(t)$ can be determined using capacity adjustment growth electric $CA^G_e(\text{t})$ and capacity adjustment electric $CA_e(t)$.

$$DAR_e(\text{t}) = CA_e(t) + CA^G_e(\text{t}) \tag{4.10}$$

The desired supply line for hydrogen $DSL^i_{hydrogen}(t)$ is given by

$$DSL^i_{hydrogen}(t) = \tau^{ED}(t) \cdot DAR^i_{hydrogen}(t). \tag{4.11}$$

The desired supply line for electric $DSL_e(t)$ is given by

$$DSL_e(t) = \tau^{ED} \cdot DAR_e(t). \tag{4.12}$$

Using the above equations, the supply line adjustment for hydrogen is computed as

$$SLA^i_{hydrogen}(t) = \frac{DSL^i_{hydrogen}(t) - SO^i_{hydrogen}(t)}{\tau^{SL}}. \tag{4.13}$$

The supply line adjustment for electric is computed as

$$SLA_e(t) = \frac{DSL_e(t) - SO_e(t)}{\tau^{SL}}. \tag{4.14}$$

The supply line adjustment growth for hydrogen then becomes

$$SLA^{G,i}_{hydrogen}(t) = SO^i_{hydrogen}(t) \cdot W^G \cdot g^{expected}(t). \tag{4.15}$$

Also, the supply line adjustment growth for electric then becomes

$$SLA_e^G(t) = SO_e(t) \cdot W^G \cdot g^{expected}(t). \tag{4.16}$$

Finally, the indicated order rate for hydrogen is then

$$IOR^i_{hydrogen}(t) = DAR^i_{hydrogen}(t) + SLA^i_{hydrogen}(t) + SLA^{G,i}_{hydrogen}(t). \tag{4.17}$$

The indicated order rate for electric aircraft is then calculated as

$$IOR_e(t) = DAR_e(t) + SLA_e(t) + SLA_e^G(t). \tag{4.18}$$

Aircraft life cycle and purchase decision

The indicated order rate calculated previously can now be used to calculate the order rates placed by airlines. This is divided into hydrogen and electric aircraft similar to the previous section. The base of these equations comes from (Kieckhäfer et al. 2018)'s overall ATS model as well as the technical report (Talwar et al. 2021a).

The two order rates for electric and hydrogen $OR_e(t)$ and $OR^i_{hydrogen}(t)$ need to be adjusted to account for aircraft that are currently out of service and are planned to return to service which is denoted by $RS(t)$. In addition, the order rate is set to zero if the previous calculation corresponds to a negative IOR, therefore, the equation becomes

$$OR^i_{hydrogen}(t) = \max\left(0, IOR^i_{hydrogen}(t) - RS^i_{hydrogen}(t)\right) \tag{4.19}$$

$$OR_e(t) = \max(0, IOR_e(t) - RS_e(t)). \tag{4.20}$$

The change of the ordered seats $SO(t)$ are determined by the order rates, delivery rates $DR(t)$ and cancelations $CX(t)$

$$\frac{d}{dt} SO^i_{hydrogen}(t) = OR^i_{hydrogen}(t) - DR^i_{hydrogen}(t) - CX^i_{hydrogen}(t), \tag{4.21}$$

$$\frac{d}{dt} SO_e(t) = OR_e(t) - DR_e(t) - CX_e(t). \tag{4.22}$$

The equation for cancelations is determined below. Over here, if the indicated order rate $IOR(t)$ is negative then the cancelations are executed. Cancelation rates are minimum of either $IOR(t)$ itself or the ordered seats $SO(t)$ divided by the duration of the cancelation τ^{CX}

$$CX^{i}_{hydrogen}(t) = \begin{cases} \min\left(-IOR^{i}_{hydrogen}(t), \frac{SO^{i}_{hydrogen}(t)}{\tau^{CX}}\right), & IOR^{i}_{hydrogen}(t) < 0 \\ 0, & else \end{cases}, \tag{4.23}$$

$$CX_{e}(t) = \begin{cases} \min\left(-IOR_{e}(t), \frac{SO_{e}(t)}{\tau^{CX}}\right), & IOR_{e}(t) < 0 \\ 0, & else \end{cases}. \tag{4.24}$$

The delivery rates $DR(t)$ is determined by dividing the number of seats ordered by the target delivery time τ^{DT} for the manufacturers of the aircraft. The function is also limited by the manufacturer's $AMC(t)$ capacity.

$$DR^{i}_{hydrogen}(t) = \min\left(\frac{SO^{i}_{hydrogen}(t)}{\tau^{DT}_{hydrogen}}, AMC^{i}_{hydorgen}(t)\right), \tag{4.25}$$

$$DR_{e}(t) = \min\left(\frac{SO_{e}(t)}{\tau^{DT}_{e}}, AMC_{e}(t)\right). \tag{4.26}$$

The aircraft operated $AU(t)$ is calculated by the addition of the delivery rate, older aircraft returning to service and subtraction of decommissioned aircraft $IS(t)$ and $RT(t)$

$$\frac{d}{dt}AU^{i}_{hydrogen}(t) = DR^{i}_{hydrogen}(t) + RS^{i}_{hydrogen}(t) - IS^{i}_{hydrogen}(t) - RT^{i}_{hydrogen}(t). \tag{4.27}$$

$$\frac{d}{dt}AU_{e}(t) = DR_{e}(t) + RS_{e}(t) - IS_{e}(t). \tag{4.28}$$

$IS(t)$ is the rate of aircraft are taken out of service. $IS(t)$ is calculated by with respect to changes in $IOR(t)$. If $IOR(t)$ is negative, it indicates that the amount of fleet in the system should be reduced. For this particular decision, there are different equations for electric and hydrogen aircraft. For hydrogen, the equation is given as

$$IS^{i}_{hydrogen}(t) = \begin{cases} \min\left(-IOR^{i}_{hydrogen}(t), \frac{AU^{i}_{hydrogen}(t)}{\tau^{ISto}}\right), & IOR^{i}_{hydrogen}(t) < 0,\ OM(t) \leq OM^{REF} \\ 0, & else \end{cases}, \tag{4.29}$$

Over here, if the operating margin $OM(t)$ falls below a reference value OM^{REF}, the $IS(t)$ becomes active. Aircraft out of service hydrogen $IS^{i}_{hydrogen}(t)$ is equal to the minimum value of aircraft in service divided by the time τ^{ISto} taken to take an aircraft out of service or indicated order rate.

The equation for the return of aircraft into service from storage is calculated by having $IOR(t)$ as a necessary condition. If $IOR(t)$ is positive, aircraft already owned by the airline will be reactivated. This rate $RS(t)$ is equal to $AS(t)$ divided by τ^{ISer}. In other words, number of parked aircraft divided by the time taken to return into service. The decision to return hydrogen aircraft from service becomes

$$RS^{i}_{hydrogen}(t) = \begin{cases} \frac{AS^{i}_{hydrogen}(t)}{\tau^{ISer}}, & IOR^{i}_{hydrogen}(t) > 0 \\ 0, & else \end{cases}. \tag{4.30}$$

The retirements of conventional aircraft from usage is calculated by dividing the average aircraft life AVG_{life} with aircraft in use $AU^{i}_{hydrogen}(t)$

$$RT^{i}_{hydrogen}(t) = \max\left(0, \frac{AU^{i}_{hydrogen}(t)}{AVG_{life}}\right). \tag{4.31}$$

Now, the storage decision for electric aircraft after the introduction of the technology is given as

$$IS_{e}(t) = \begin{cases} \min\left(-IOR_{e}(t), \frac{AU_{e}(t)}{\tau^{ISto}}\right), & IOR_{e}(t) < 0, OM(t) \leq OM^{REF} \\ 0, & else \end{cases}. \tag{4.32}$$

The return into storage for the electric aircraft is

$$RS_{e}(t) = \begin{cases} \frac{AS_{e}(t)}{\tau^{ISer}}, & IOR_{e}(t) > 0 \\ 0, & else \end{cases}. \tag{4.33}$$

4.2.3.2 Electric Aircraft Costs

The battery charging cost $BC(t)$ for electric aircraft is determined by multiplying the available seat kilometer offered by electric aircraft $ASK_{e}(t)$ and the electricity consumption per seat kilometer $BEC^{ASK}(t)$ and the electricity price $ER(t)$. In this equation, the electricity price is subvented by using an electricity policy denoted as $Sub^{ER}(t)$

$$BC(t) = BEC^{ASK}(t) \cdot ASK_e(t) \cdot \left(ER(t) - Sub^{ER}(t)\right). \tag{4.34}$$

The per unit cost of electric aircraft can be determined by the ownership costs $OC_e^{seat \cdot km}(t)$ and the operational costs $OOC^{ASK}(t)$, inclusive of the battery charging cost $BC(t)$ divided by available seat kilometers of electric aircraft $ASK_e(t)$.

$$UC_e(t) = OC_e^{seat \cdot km}(t) + OOC^{ASK}(t) + \frac{BC(t)}{ASK_e(t)}. \tag{4.35}$$

Cost comparison of conventional and electric aircraft

Regarding the decision of the airline to choose one of the two available technologies i.e., conventional and electric, the unit cost for operation for technologies are compared. The relative difference in the operating cost $IC^{Diff}(t)$ is calculated as follows:

$$IC^{Diff}(t) = IC_e(t) - IC_{conv}(t) \tag{4.36}$$

In this equation the introduction cost of operation $IC_{conv}(t)$ for conventional is determined by the sum of fuel costs, operational costs $OOC^{ASK}(t)$ and ownership costs $OC_{conv}^{seat \cdot km}(t)$ per unit. $FU^{NA}(t)$ is the fuel consumption of new aircraft, alternative fuel share is denoted by $AFS(t)$, price of alternative fuel is denoted by $AFP(t)$ and $CFP(t)$ is the price of conventional kerosene.

$$\begin{aligned} IC_{conv}(t) = & FU^{NA}(t) \cdot AFS(t) \cdot AFP(t) + FU^{NA}(t) \cdot (1 - AFS(t)) \\ & \cdot \left(CFP(t) + Tax^{fuel}(t)\right) + OOC^{ASK}(t) + OC_{conv}^{seat \cdot km}(t). \end{aligned} \tag{4.37}$$

Ownership cost per seat-kilometer $OC_{conv}^{seat \cdot km}(t)$ is obtained by dividing the ownership cost per new seat by the average annual flight distance of a seat $distance_{conv}^{seat}$

$$OC_{conv}^{seat \cdot km}(t) = \frac{OC_{conv}^{year}(t)}{distance_{conv}^{seat}}. \tag{4.38}$$

The operating cost for the introduction of electric aircraft $IC_e(t)$ consists of electricity costs due to the charging of batteries denoted by $BEC^{ASK}(t)$ multiplied by electricity price $ER(t)$. As mentioned above $Sub^{ER}(t)$ is the electricity price subvention offered by the government.

$$IC_e(t) = BEC^{ASK}(t) \cdot \left(ER(t) - Sub^{ER}(t)\right) + OOC^{ASK}(t) + OC_e^{seat \cdot km}(t). \tag{4.39}$$

The ownership costs per seat-kilometer for electric aircraft are kept analogously to the conventional ones

$$OC_e^{seat \cdot km}(t) = \frac{OC_e^{year}(t)}{distance_e^{seat}}. \tag{4.40}$$

4.2.3.3 Aircraft Manufacturers

The base of the equations for the aircraft manufacturing sub-section is taken from (Kieckhäfer et al. 2018). The targeted manufacturing capacity $TMC(t)$ is determined using the open orders denoted by $SO^i_{Conv}(t)$ divided by the delivery time τ^{DT} in which the orders are to be executed. The targeted manufacturing capacity $TMC^i_{hydrogen}(t)$ is calculated as

$$TMC^i_{hydrogen}(t) = \max\left(0, \frac{SO^i_{hydrogen}(t)}{\tau^{DT}}\right). \tag{4.41}$$

The actual manufacturing capacity $AMC^i_{hydrogen}(t)$ is equal to the change in manufacturing capacity.

$$\frac{d}{dt} AMC^i_{hydrogen}(t) = AMC^i_{hydrogen,change}(t), \tag{4.42}$$

where the change in manufacturing capacity $AMC^i_{Hydrogen,Change}(t)$ is given by

$$AMC^i_{hydrogen,change}(t) = \frac{TMC^i_{hydrogen}(t) - AMC^i_{hydrogen}(t)}{\tau^{MC}}. \tag{4.43}$$

In the case of electric aircraft, the target manufacturing capacity electric $TMC_e(t)$ is given by

$$TMC_e(t) = \frac{SO_e(t)}{\tau_e^{DT}}, \tag{4.44}$$

$$\frac{d}{dt}AMC_e(t) = \begin{cases} \frac{TMC_e(t)-AMC_e(t)}{\tau_e^{MC}}, & t \geq t_{short} \\ 0, & else \end{cases}. \tag{4.45}$$

Furthermore, account is taken of the fact that the manufacturing process of the new type of aircraft has just been started and therefore the capacity cannot be increased as quickly as conventional aircraft. For this reason, improvement rates are included in the model. The Improvement in the capacity adjustment time for electric τ_e^{MC} is calculated by multiplying it with the time taken at the initial level $\tau_e^{MC,\ t0}$ and improvements $I(t)$.

$$\tau_e^{MC} = \tau_e^{MC,t0} \cdot I(t). \tag{4.46}$$

Further clarifications regarding the improvements are discussed in the technical report (Talwar et al. 2021b).

4.2.3.4 Fuel Producers and Energy Requirements

Since jet fuel can be replaced by sustainable aviation fuels, their introduction in the model is done by multiplying the quota by the annual fuel consumption. The annual fuel consumption $FU_{ANN}^{i,Conv}(t)$ for each segment for conventional is calculated by multiplying fuel consumption per available seat kilometer by available seat kilometer (ASK),

$$FU_{ANN}^{i,Conv}(t) = FU_{ASK}^{i,Conv}(t) \cdot ASK_{Conv}(t). \tag{4.47}$$

The target amount of SAF is then calculated by

$$Target\ SAF^i(t) = FU_{ANN}^{i,Conv}(t) \cdot Quota\ SAF^i. \tag{4.48}$$

Here, $Quota\ SAF^i$ is the percentage share of SAF planned for the system (Mckinsey 2020; IATA 2021).

Hydrogen fuel production

In the model, hydrogen fuel producers track the amount of hydrogen fuel being consumed by the airlines to adjust for their capacities. The initial level of hydrogen availability when hydrogen based aircraft are introduced in the year 2035 is taken from the clean sky report (Fuel Cells and Hydrogen Joint Undertaking 2020). The annual hydrogen fuel consumption in the system is given by

$$HFU^i_{ANN}(t) = HFU^i_{ASK_{hydrogen}}(t) \cdot ASK^i_{hydrogen}(t). \tag{4.49}$$

Here, $HFU^i_{ANN}(t)$ is the annual fuel consumption of hydrogen, $HFU^i_{ASK_{hydrogen}}(t)$ is the hydrogen fuel consumption per available seat kilometer and $ASK^i_{hydrogen}(t)$ is the available seat kilometers for the hydrogen fleet.

The amount of hydrogen fuel targeted to be produced by the fuel producers is equal to the annual fuel consumption by the airlines. By setting this as a target the fuel producers can be more certain about the capacity changes,

$$HFU^i_{Tar}(t) = HFU^i_{ANN}(t). \tag{4.50}$$

The hydrogen fuel production capacity is given by

$$\frac{d}{dt}HFU^i_{capacity}(t) = HFU^i_{change}(t). \tag{4.51}$$

here $HFU^i_{change}(t)$ the change in production is given by

$$HFU^i_{change}(t) = \frac{HFU^i_{adj}(t)}{\tau^h}. \tag{4.52}$$

Here, $HFU^i_{adj}(t)$ is the adjustment for the capacity and τ^h is the time required by the fuel producers to adjust the capacity. The adjustment of the capacity is calculated by subtracting the existing capacity from the targeted capacity,

$$HFU^i_{adj}(t) = HFU^i_{Tar}(t) - HFU^i_{capacity}(t) \tag{4.53}$$

Equations (4.50) to (4.53) create a goal seeking behavior where an increase in the usage of hydrogen aircraft increases the fuel usage. The increase in fuel usage then prompts the fuel producers to adjust their capacities, thereby reaching the fuel production goals. The share of green hydrogen in the model is calculated by multiplying the hydrogen capacity by the green hydrogen share,

$$HFU^i_{green}(t) = HFU^i_{capacity}(t) \cdot GHFU. \tag{4.54}$$

Over here, $GHFU$ is the green hydrogen fuel share taken externally from the different reports that, according to current developments predict the possible share of green hydrogen by 2050 (IEA 2021b).

Green electricity requirements for green hydrogen

The green electricity required for producing green hydrogen is calculated in a similar goal seeking equation setting as in Equations (4.50)–(4.53) . The target green electricity requirement is calculated by multiplying the green hydrogen capacity with the electricity required per kg of hydrogen to produce green hydrogen,

$$ELC^i_{target}(t) = HFU^i_{green}(t) \cdot ELC^i_{req}, \tag{4.55}$$

where $ELC^i_{target}(t)$ is the target amount of green electricity required and ELC^i_{req} is the electricity required per kilogram of hydrogen. This value is taken from the literature according to the current capabilities to produce green hydrogen (Kakoulaki et al. 2021).

The green electricity availability for the green hydrogen required by aviation is represented in the model by

$$\frac{d}{dt}ELC^i_{availability}(t) = ELC^i_{change}(t), \tag{4.56}$$

where change in availability $ELC^i_{change}(t)$ is given as

$$ELC^i_{change}(t) = \frac{ELC^i_{adj}(t)}{\tau^e}. \tag{4.57}$$

$ELC^i_{adj}(t)$ is the adjustment for the capacity and τ^e is the time required to establish green electricity capacity. Finally, the relationship between adjustment, target and availability of green electricity is given by

$$ELC^i_{adj}(t) = ELC^i_{target}(t) - ELC^i_{availability}(t). \tag{4.58}$$

Sustainable aviation fuel production

In the case of sustainable aviation fuels, the target volume to be produced is equal to the target amount SAF planned in the system,

$$SAF^i_{target\ volume}(t) = Target\ SAF^i(t), \tag{4.59}$$

where $SAF^i_{target\ volume}(t)$ is the target production volume of SAF and $Target\ SAF^i(t)$ is taken over from (4.48). Here, i represents the range segment.

Once the target volume is set, the production capacity is represented as

$$\frac{d}{dt}SAF^{i}_{capacity}(t) = SAF^{i}_{change}(t), \tag{4.60}$$

where the change in production $SAF^{i}_{change}(t)$ is given by

$$SAF^{i}_{change}(t) = \frac{SAF^{i}_{adj}(t)}{\tau^{SAF}}. \tag{4.61}$$

Here, $SAF^{i}_{adj}(t)$ is the adjustment for the capacity and τ^{SAF} is the time required by the fuel producers to adjust the capacity. Finally, the adjustment required is calculated by subtracting the current capacity from the targeted volume,

$$SAF^{i}_{adj}(t) = SAF^{i}_{targetvolume}(t) - SAF^{i}_{capacity}(t). \tag{4.62}$$

Since different types of SAFs are considered in the model, the separation of the different pathways from the main capacity is done by multiplying the capacity by the share of each type of SAF. This is done by introducing subscripts in the model,

$$SAF^{i,j}_{capacity,PtL}(t) = SAF^{i}_{capacity}(t) \cdot SAF^{j}_{type}. \tag{4.63}$$

Here, j is the type of SAF, where $j = 1$ is HEFA, $j = 2$ is FT synthesis, $j = 3$ is AtJ and $j = 4$ is PtL. As mentioned in the literature review section, in this thesis focus is more on the PtL type of SAF as they require green electricity and hydrogen. SAF^{j}_{type} is the share of each type of SAF that is expected to develop over the years according to the reports (Airbus 2018; ATAG 2020b; ICCT 2021).

Hydrogen requirement for PtL type SAF

The equation for the hydrogen required for the PtL type of SAF is given by

$$HFU^{i,j}_{target,PtL}(t) = SAF^{i,j}_{capacity,PtL}(t) \cdot Hydrogen^{i,j}_{req,PtL}(t). \tag{4.64}$$

$HFU^{i,j}_{target,PtL}(t)$ is the target hydrogen fuel requirement for the PtL type SAF and $Hydrogen^{ij}_{req,PtL}(t)$ is the hydrogen required to produce one kilogram of SAF from the PtL pathway. This is taken externally from the different reports in light of the current developments (ATAG 2020b).

The hydrogen requirement capacity change for SAF production through PtL pathway is

$$\frac{d}{dt}HFU^{i,j}_{capacity,PtL}(t) = HFU^{i,j}_{change,PtL}(t), \tag{4.65}$$

where the change in production $HFU^{i,j}_{change,PtL}(t)$ is given by

$$HFU^{i,j}_{change,PtL}(t) = \frac{HFU^{i}_{adj,PtL}(t)}{\tau^h}. \tag{4.66}$$

Here, $HFU^{i,j}_{adj,PtL}(t)$ is the adjustment for the capacity and τ^h is the time required by the fuel producers to adjust the capacity. The time taken to adjust the capacity for PtL is similar to the time taken in hydrogen fuel production capacity.

$$HFU^{i,j}_{adj,PtL}(t) = HFU^{i}_{Tar,PtL}(t) - HFU^{i,j}_{capacity,PtL}(t). \tag{4.67}$$

Similar to (4.54) the share of green hydrogen for the PtL type of SAF is calculated by multiplying the current capacity $HFU^{i,j}_{capacity,PtL}(t)$ by the potential future green hydrogen developments.

$$HFU^{i,j}_{green,PtL}(t) = HFU^{i,j}_{capacity,PtL}(t) \cdot GHFU^{i,j}. \tag{4.68}$$

Green electricity requirement for green hydrogen for PtL type SAF

The equations for green electricity requirement for producing the hydrogen for the PtL route are kept similar to the ones used for green hydrogen fuel. The only difference is the subscript for the PtL type of SAF. The target green electricity requirement for PtL SAF is calculated by multiplying the green hydrogen capacity with the electricity required per kg of hydrogen to produce green hydrogen.

$$ELC^{i,j}_{target,PtL}(t) = HFU^{i,j}_{green,PtL}(t) \cdot ELC^{i,j}_{req,PtL}, \tag{4.69}$$

In (4.69), $ELC^{i,j}_{target,PtL}$ is the electricity required per kilogram of hydrogen. This value is taken from the literature and reports according to the current capabilities to produce green hydrogen through various methods (Kakoulaki et al. 2021). The green electricity availability for the green hydrogen for SAF from PtL pathway is represented by

$$\frac{d}{dt}ELC^{i,j}_{availability,PtL}(t) = ELC^{i,j}_{change,PtL}(t), \tag{4.70}$$

where the change in availability $ELC^{i,j}_{change,PtL}(t)$ is given as

$$ELC^{i,j}_{change,PtL}(t) = \frac{ELC^{i,j}_{adj,PtL}(t)}{\tau^e}. \tag{4.71}$$

τ^e is the time required by the electricity producers to adjust for capacity similar to (4.57). Finally, the relationship between adjustment, target and availability of green electricity is

$$ELC^{i,j}_{adj,PtL}(t) = ELC^{i,j}_{target,PtL}(t) - ELC^{i,j}_{availability,PtL}(t). \tag{4.72}$$

4.2.3.5 Policies, Implications and Improvements

The previous models did not contain of different types of policies that can encourage the adoption of new technology. In the extended model, jet fuel tax, electricity subsidy, seat tax and government restriction policy are added. The explanation of these policies is as follows:

Tax and subsidy

The electricity subsidies are implemented by providing subsidies on the electricity price used to charge the batteries of the aircraft. This subsidy option was included in the computation of the battery costs described in (4.34) as electricity subsidy $Sub^{ER}(t)$.

$$[(34) : BC(t) = BEC^{ASK}(t) \cdot ASK_e(t) \cdot \left(ER(t) - Sub^{ER}(t)\right)].$$

Here, the subsidy is an exogenous parameter that can be given a value that will determine the extent to which the subsidy is provided. In addition to subsidies, the jet fuel tax is also included in the model as a political incentive measure to promote the usage of electric technology. As discussed in Sect. 2.5, to decrease the attractiveness of conventional aircraft, jet fuel can be implemented. It can make conventional aircraft less attractive (González and Hosoda 2016; Fukui and Miyoshi 2017). This is implemented by adding a tax $Tax^{fuel}(t)$ on the usage of conventional kerosene/jet fuel. $FC(t)$ is the fuel cost for airlines, this consists of the costs of alternative and conventional fuel. In this equation, the share of alternative fuels is denoted by $AFU^{ANN}(t)$ and the total jet fuel consumption is denoted by $FU^{ANN}(t)$. Therefore, the equation for fuel cost is given as follows

$$FC(t) = AFU^{ANN}(t) \cdot AFP(t) + \left(FU^{ANN}(t) - AFU^{ANN}(t)\right) \cdot \left(CFP(t) + Tax^{fuel}(t)\right). \quad (4.73)$$

Government fleet restriction policy and penalty costs

The government fleet restriction policy refers to a scenario in which the government dictates to airlines that a specific portion of their fleet must be operated using electric aircraft. When the government takes this decision, then the airlines behave accordingly in order to not lose their profits.

First, the airlines calculate the amount of conventional fleet that has to be put in storage so that the percentage of electric fleet can be replaced by it. This decision is explained in the equations as follows

$$AC_{conv}^{store}(t) = \frac{Target_{conv}(t)}{Flight\ distance\ per\ seat_{conv}}, \quad (4.74)$$

Here, $AC_{conv}^{store}(t)$ is the amount of conventional aircraft that need to go in storage so that they can be replaced with battery electric aircraft. $Flight\ distance\ per\ seat_{conv}$ is the average amount of distance traveled by a seat operated by conventional aircraft. In the equation, $Target_{conv}(t)$ is given by

$$Target_{conv} = \begin{cases} Max(ASK_{conv}(t) - OB_{conv}(t), 0), & t > t_{policy}, \\ 0, & else \end{cases} \quad (4.75)$$

Here, the $OB_{conv}(t)$ is the objective for the airlines calculated as follows:

$$OB_{conv}(t) = ASK_{conv}(t) \cdot Multiplier_{conv} \quad (4.76)$$

The here $Multiplier_{conv}$ reflects the mandatory share of the electric fleet set by the government. In other words, a higher value would mean a more stringent policy by the government to achieve higher percentage of electric aircraft in the system. Therefore, the multiplier for conventional would also take a higher value. These particular seats will be then put into storage by the conditions set by initiating storage given as

$$IS_{conv}(t) = \begin{cases} AC_{conv}^{store}(t), & IC^{diff}(t) > 0 \\ \min\left(-IOR_{conv}(t), \frac{AU_{conv}(t)}{\tau^{ISto}}\right), & IOR_{conv}(t) < 0 \text{ and } DC_e(t) > 0 \\ 0, \text{ else} \end{cases}, \tag{4.77}$$

Furthermore, these aircraft will only return to the service under the condition as follows

$$RS_{conv}(t) = \begin{cases} 0, & IC^{diff}(t) < 0, \\ \frac{AS_{conv}(t)}{\tau^{ISer}}, & IC^{diff}(t) > 0, \quad Multiplier_{conv} < 0. \\ 0, & else \end{cases} \tag{4.78}$$

After storing the conventional aircraft that need to be replaced by electric aircraft, the airlines make changes in their fleet ordering given by the equation

$$DC_e^{policy}(t) = \begin{cases} DC(t) \cdot Multiplier^{policy}, & t > t_{short} \\ 0, & else \end{cases} \tag{4.79}$$

$Multiplier^{policy}$ is the percentage share of electric aircraft mandated by the government imposed on the airline operators. The $DC_{conv}^{policy}(t)$ is determined by subtracting the above calculated fleet policy ordering from the actual desired capacity

$$DC_{conv}^{policy}(t) = DC(t) - DC_e^{policy}(t) \tag{4.80}$$

With the help of this mechanism, the airlines increase the desired capacity for an electric fleet and reduce their conventional fleet as explained in Equation (4.81)

$$DC_e(t) = \begin{cases} 0, & t < t_{short} \\ DC(t), & t > t_{short} \text{ and } IC_{per}^{diff}(t) \leq 0 \\ DC_e^{policy}(t), & t > t_{short}, \ IC_{per}^{diff}(t) \text{ and } t > t_{policy} \\ 0, & else \end{cases} \tag{4.81}$$

The above equations suggest that, if the government fleet restriction policy had not been introduced then the airlines after the introduction of the electric aircraft make their decisions to order more electric aircraft solely based on the introduction cost difference $IC^{diff}(t)$. The fleet policy is modeled in a way that regardless of the cost competitiveness of electric aircraft, the airlines have to increase their usage over time.

The stored conventional aircraft would incur costs for the airlines as the aircraft that are not being used but in storage have to be maintained and incur parking fees. As a result of this, the airlines lose profit that they otherwise would have earned if the aircraft had been operational. This type of loss is included as the opportunity cost lost and is added to the unit cost for conventional aircraft $UC_{conv}(t)$. The government fleet policy is only focused on increasing the adoption through mandate but another discouragement for conventional fleet done with the help of a seat tax. The seat tax policy is a situation in which the government applies a tax on every seat that is operated by using conventional aircraft. The computation of both above mentioned aspects is given as follows

$$UC_{conv}(t) = \begin{cases} \frac{FC(t)+OOC^{ASK}(t)+Storage^{per}(t)}{ASK_{conv}(t)} + OC_{conv}(t) + Penalty^{per}(t), & t > t_{short} \\ \frac{FC(t)+OOC^{ASK}(t)}{ASK_{conv}(t)} + OC_{seat}(t), & else \end{cases} \tag{4.82}$$

Over here, $Penalty^{per}(t)$ is the perceived penalty costs and $Storage^{per}(t)$ is the perceived storage costs. These variables are denoted consist of a delay where the airlines need some time to compute the costs. The computation of $Storage^{per}(t)$ is given as follows

$$Profit^{TTL}(t) = Revenue^{TTL}(t) - Costs^{TTL}(t) \tag{4.83}$$

Over here, $Revenue^{TTL}$ is the total revenue earned by airlines in a calendar year and $Costs^{TTL}(t)$ is the total expenditures of airlines to operate the fleet in a calendar year. With the help of the computed profit, the profit per seat $Profit_{seat}(t)$ can be determined by dividing it by the total fleet denoted by $ASK^{TTL}(t)$.

$$Profit_{seat}(t) = \frac{Profit^{TTL}(t)}{ASK^{TTL}(t)} \tag{4.84}$$

From the equation above, the opportunity cost lost due to aircraft in storage can be determined.

$$Storage^{TTL}(t) = AS_{conv}(t) \cdot Profit_{seat}(t) \tag{4.85}$$

In the above equation, few delays are introduced and then the $Storage^{TTL}(t)$ becomes $Storage^{per}(t)$.

Regarding the penalty cost, the penalty fee per year $Penalty_{year}(t)$ depends on the introduction of government fleet policy and if the share of the operated electric fleet is less than the specified mandate.

$$Penalty_{year}(t) = \begin{cases} Penalty\,fee\,rate, & t > t_{policy} \quad , \quad ASK_e^{ratio} < Multiplier_{policy} \\ 0, & else \end{cases} \tag{4.86}$$

Over here, *Penalty fee rate* is calculated with the help of the Monte Carlo sensitivity simulation technique in which various values of these variables are tested in the model. Furthermore, when some delays are introduced in the penalty per year $Penalty_{year}(t)$ becomes $Penalty^{per}(t)$.

Improvements and learnings

The model considers technological and structural improvements that help in cost reductions. This is applied to hydrogen and electric airplanes analogously.

The speed at which improvement $I(t)$ takes place depends on the constant rate of improvement IR and time τ^{IR}

$$\frac{d}{dt}I(t) = I(t) \cdot \frac{IR}{\tau^{IR}}. \tag{4.87}$$

$I(t)$ has an initial value of 1 and decreases with time. Thus, the improvement $I(t)$ can be used as a technological improvement by multiplying it to corresponding variables to reduce their values over time. For conventional aircraft, this takes place at the fuel consumption of a new aircraft given as follows

$$FU^{NA}(t) = FU^{NA}(t) \cdot I(t). \tag{4.88}$$

In terms of cost of electric aircraft, ownership cost per seat is equal to initial ownership cost value $OC_e^{initial}$ per seat kilometer multiplied by the average lifetime of an aircraft AC_e^L and the flight distance per seat $distance_e^{seat}$. These costs are incurred only when the electric aircraft are introduced in the model. Here, the improvement effect $I(t)$ is directly multiplied. In this case, the improvements are assumed to take effect in the form of economies of scale which reduce the ownership costs over time.

$$OC_e(t) = \begin{cases} AC_e^L \cdot distance_e^{seat} \cdot OC_e^{initial} \cdot I(t) \ , & t \geq t_{short} \\ 0 \ , & else \end{cases} \tag{4.89}$$

For conventional ownership costs, instead of using improvements, the consumer price inflation index (CPI) is used to increase the ownership cost over time as improvements in conventional aircraft are only considered from the fuel consumption side.

$$OC_{conv}(t) = AC^{L}_{conv} \cdot distance^{seat}_{conv} \cdot OC^{initial}_{conv} \cdot CPI(t) \tag{4.90}$$

Similarly, for hydrogen, the improvements for the new aircraft are calculated as follows

$$FU^{NA}_{hydrogen}(t) = FU^{NA}_{hydrogen}(t) \cdot I(t). \tag{4.91}$$

In terms of improvement in costs for hydrogen aircraft, the improvements are used in the calculation of the other operating cost per ASK variable $OOC^{ASK}_{Hydrogen}(t)$ given as follows

$$OOC^{ASK}_{Hydrogen}(t) = \frac{OOC_{hydrogen}}{AU_{hydrogen}} \cdot CPI \cdot I(t). \tag{4.92}$$

Here, $OOC_{hydrogen}$ is the other operating cost for hydrogen and CPI is the consumer price inflation.

4.3 Model Testing and Validation

In system dynamics methodology, there are many ways to conduct validation tests for a model. The two most important measures are the model structural test and the model behavioral test (Sterman 2000). In this section, the structural and behavioral tests done on the extended model are showcased. In the first part of the section, the model structural tests are done where the model is tested against historical data as well as for the dimensional, integration and extreme condition tests. In the second part, the model behavioral tests in the form of extreme policy tests are done to generate confidence in the model.

4.3.1 Model Structural Tests

To develop confidence in the structure of the model, various validation tests need to be conducted in accordance with the system dynamics literature (Forrester and Senge 1980; Sterman 2000). One of the most important structural tests to develop confidence in a model according to system dynamics practice are namely: historical data tests, dimensional and integration tests and extreme conditions tests. The results of these tests are explained below.

Table 4.2 Model validation: splitting of the mean squared error

Variable	Splitting of the MSE		
	Bias	Unequal variation	Unequal covariance
Aircraft in use	0.089	0.064	0.847
Airline fares	0.006	0.094	0.901
Air travel demand	0.035	0.341	0.623
Annual aviation CO_2 emissions	0.000	0.035	0.965
Load factor	0.002	0.452	0.546
Operating margin	0.302	0.006	0.692

4.3.1.1 Testing Against Historical Data

In the system dynamics literature, testing the system dynamics model against historical data is considered to be a valuable method to develop confidence in the model. Testing a model against historical data not only helps in uncovering the flaws in the structure and assumptions of the model but also in assessing the point-by-point error measure of the model (Sterman 2000). To do this, the Theil inequality index (Theil 1966) for the six main variables of the model: aircraft in operation, ticket prices, demand, annual CO_2 emissions, seat load factor and operating margin was used. Due to data constraints, the values are only tested for the period 1991–2016. Table 4.2 shows the splitting of the mean squared errors (MSE) in unequal bias, unequal variation and unequal covariance as testing measures. It can be observed that all variables except the operating margin represent the historical data with a very low bias and unequal variation. The graphical representation of the historical data against the simulated data in Fig. 4.12 suggests the same conclusions. The low bias and unequal variation in most of the variables indicate that the model is fairly good at replicating real world values and is free from systematic errors (Sterman 2000; Kapmeier and Gonçalves 2018; Kieckhäfer et al. 2018).

In the case of the operating margin, the main reason for the higher bias could be the high complexity of revenue management and price setting mechanism in the aviation sector. This process, including artificial intelligence and other techniques, used by modern airline companies to calculate ticket prices, is difficult to simplify in a system dynamics differential equation framework. Since the goal of this study is not in the direction of revenue management, this particular discrepancy in the model can be accepted.

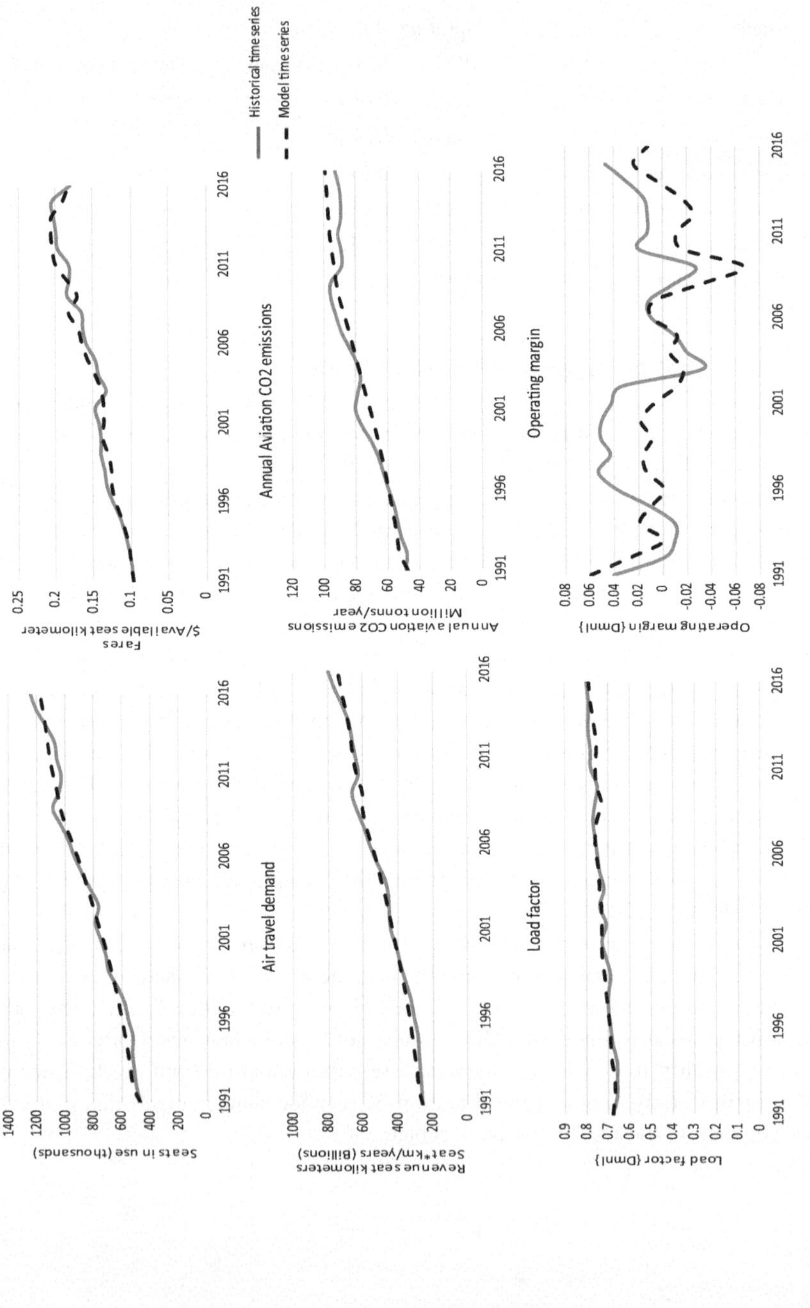

Fig. 4.12 Comparison of historical and simulated data

4.3.1.2 Dimensional, Integration and Extreme Condition Tests

To complete the validation process and develop more confidence in the model, extreme condition tests, dimensional consistency tests and integration error tests were adopted (Sterman 2000). To check for dimensional consistencies and integration errors in the model, *Vensim DSS version 8.0.8* software was used. For this, the software automatically conducts the tests and provides the results. In this regard, the model was found to be dimensionally consistent and free of any integration errors. The test results for dimensional consistency suggested that all the units used in the model are balanced. In other words, the model was dimensionally consistent throughout. On the other hand, the results of the integration check suggested that the model was free from any integration errors. For the extreme condition tests, a time delay variable was given an extreme value in the presence of the introduction of a new technology in the system. This was done to test the robustness of the extended model with respect to the extensions done in the model regarding the introduction of new technologies. In this context, the variable chosen is the time to adjust the manufacturing capacity for hydrogen. The share of hydrogen technology on the overall level was set to 20% to test for extreme conditions using the above variable. For the extreme condition test, two cases were chosen, first, the time to adjust manufacturing capacity for hydrogen was set to 50 years to account for unrealistic time delays.

Second, the time to adjust for manufacturing capacity hydrogen was set to 0.1 years to account for unrealistic fast production times. The variables aircraft in use hydrogen and aircraft manufacturing capacity hydrogen were chosen to check for plausible behavior. These two variables were chosen to check for plausible behavior, because these are the most central variables, and are directly affected by the exogenous time delay variable. Furthermore, any implausible behavior shown by the model would reflect directly in these two variables with respect to the extreme conditions in the time delay variable.

Figure 4.13 and Figure 4.14 showcase the results of the extreme condition test. It can be observed from the figures that the model is showing plausible behavior under extreme conditions. In the case when the time variable was set to 50 years, the share of hydrogen aircraft in 2050 was found to be only 9%. The increase in growth for the manufacturing capacity and aircraft in use is observed to be slow. In the case when the variable is set to 0.1 years the share of hydrogen aircraft quickly reaches the goal of 20%. Also the manufacturing capacity and aircraft in use variable increase rapidly as compared to the first case. This shows that the model showcases plausible behavior even in extreme conditions by showcasing a rapid increase when the time delay variable is set lower and slower growth when the time delay is set higher.

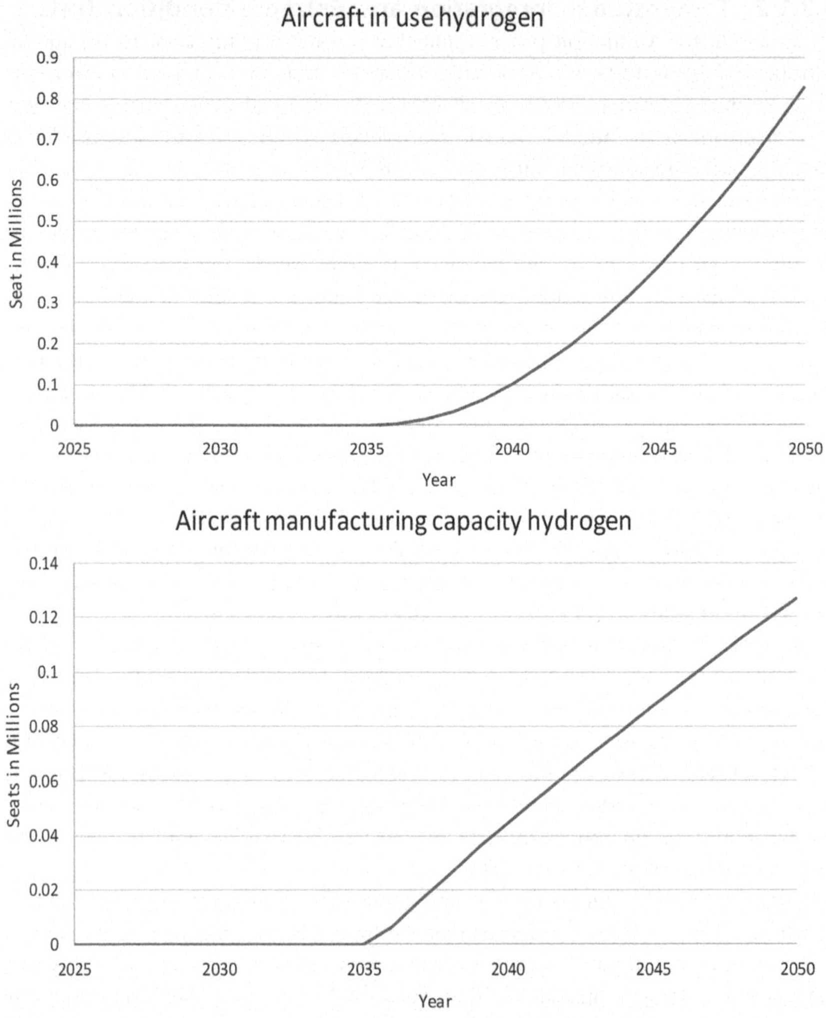

Fig. 4.13 Extreme condition test: 20% hydrogen fleet and 50 years aircraft manufacturing time delay

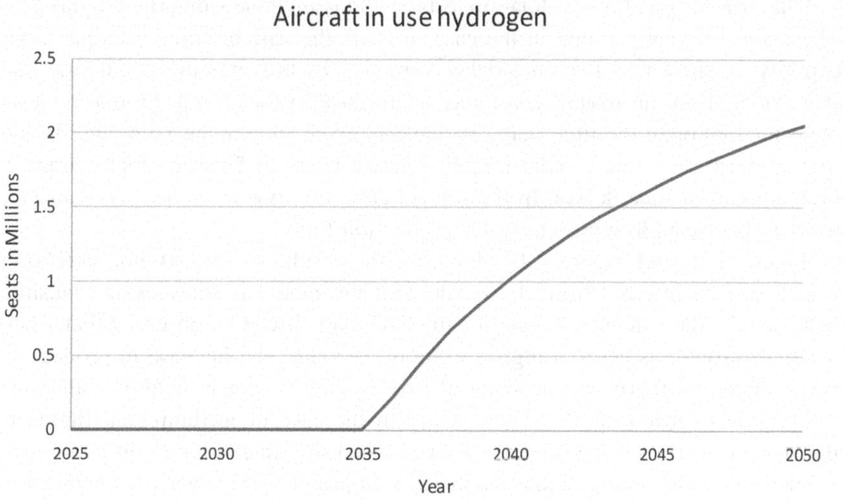

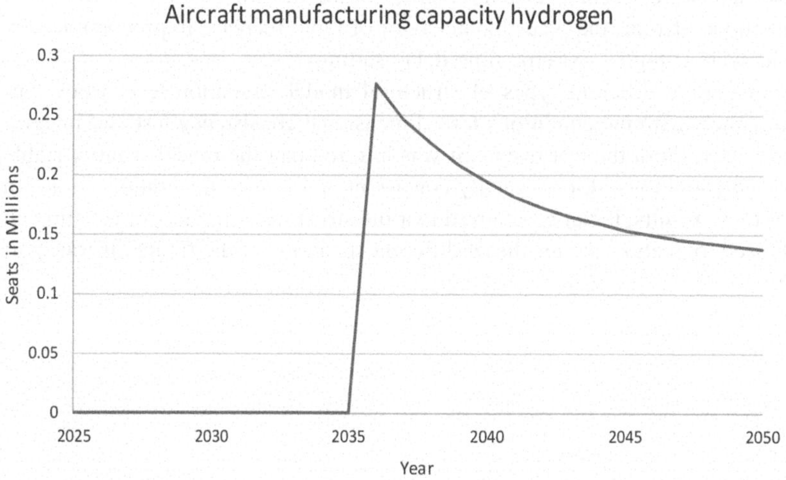

Fig. 4.14 Extreme condition test: 20% hydrogen fleet and 0.1 year aircraft manufacturing time delay

The extreme condition validation tests for SAF are done similarly to hydrogen. The main difference is that in the case of SAF, the variable time to adjust SAF capacity is chosen as the time delay variable. To test extreme conditions, the share of SAF on an overall level was set to 20% in the model. Similar to tests done for hydrogen the time delay variable is given to extreme scenarios. In the first scenario, the time to adjust SAF capacity is set to 50 years representing a high amount of time delays. In the second case, the time to adjust SAF capacity is set to 0.1 years to represent faster production times.

Figure 4.15 and Figure 4.16 showcase the results of the extreme condition test. It can be inferred from the results that the model is showcasing plausible behavior for the extensions done for the SAF part. It can be observed that when the time variable is set to a higher value of 50 years, the increase in production takes longer. Furthermore, the share of SAF by flight segment in almost all cases is not able to reach the 20% goal. Only in the case of medium-haul flights is the share near the goal. In the other case, when the time delay is set to a lower number, it can be observed that the increase in production capacity is exponential and the targets are reached relatively earlier in the model. This goes on to show that the model depicts the plausible behavior of rapid increase in production with lower as well as higher extreme time delay settings.

Overall, three different types of structural model validation tests were conducted namely extreme conditions test, dimensional consistency test and integration error test. From the test results, it was inferred that the model is fairly stable in extreme conditions, dimensionally consistent and free of integration errors. In light of these results, it can be inferred that the structure of the model is valid and can be used to analyze the results of different scenarios of the future air transport system.

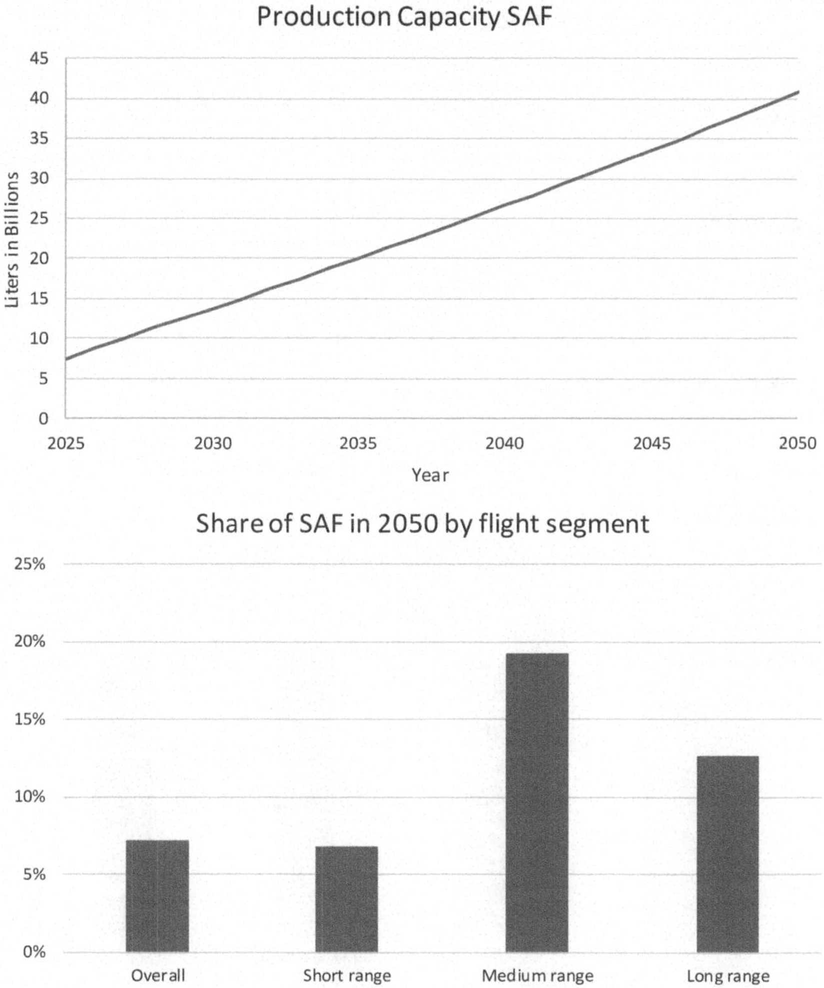

Fig. 4.15 Extreme condition test: 20% SAF and 50 years fuel production time delay

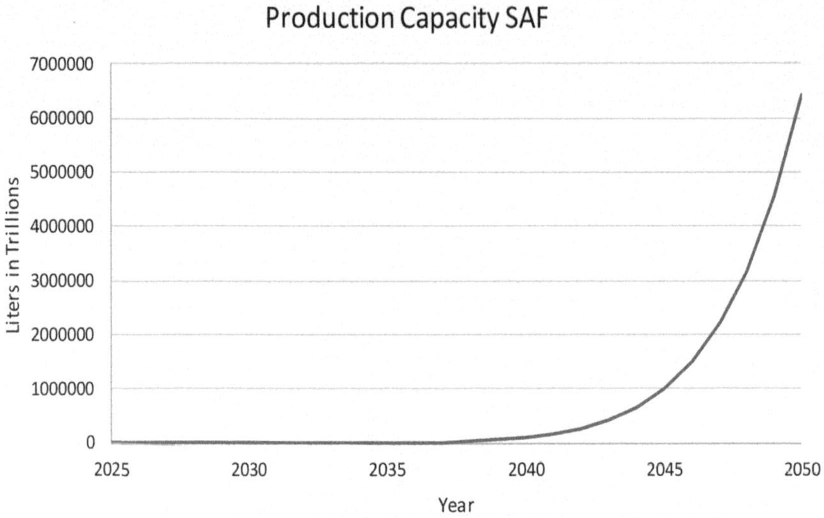

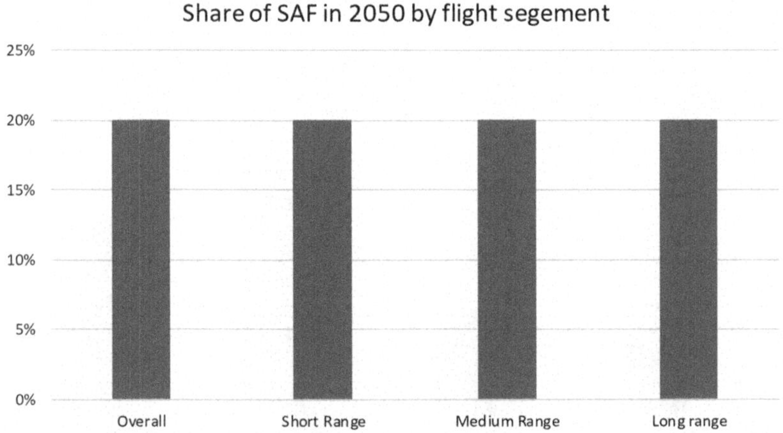

Fig. 4.16 Extreme condition test: 20% SAF and 0.1 years fuel production time delay

4.3.2 Model Behavioral Tests

In system dynamics methodology, model behavioral tests consists of tests like extreme policy tests and parameter sensitivity tests. The behavioral test results conducted for the extended model are discussed below.

4.3.2.1 Extreme Policy Tests

Conducting rigorous extreme policy is essential to ensure the confidence and validity of the behavior of a model. Extreme policy tests can be similar to surprise behavior tests where existing real system behavior is recognized only upon modeling and serves as a crucial test for unveiling unnoticed intricacies. This type of testing allows to refine and improve our model by aligning it more closely with the intricate realities of the system. Additionally, employing extreme policy tests, characterized by the introduction of radical policies into the model, serves as a litmus test for the model's consistency under unconventional conditions. These tests push the boundaries of the model, which in turn helps in gauging its robustness and assess its capacity to withstand extreme scenarios.

The model was tested for various extreme policies, e.g., a 90% fleet restriction policy or very high subsidies, to check if it shows plausible behavior in an extreme condition. The possibility of implementing such policies, in reality, is highly unlikely, but it serves as a good validation practice. For instance, suppose the government would force airlines to operate 90% of their fleet with electric aircraft after their introduction. An example of the model behavior in this case is given in Fig. 4.17. The variables ASK conventional and seats on order conventional are chosen to be observed for any irregularities because these are the most central and important variables in their respective part of the model. It can be seen in the figure that no variable takes on a negative value after the introduction of the policy in the year, 2038. On the contrary, it can be observed that the conventional fleet is reduced to meet the fleet restriction policy requirements. This suggests that in extreme conditions, the selected model variables are showing plausible behavior (Sterman 2000).

Figure 4.18 depicts the variables for the 95% fleet restriction policy in the model. It can be observed in the figure that the model variables are behaving plausibly. The only major difference is that the increase in the ordered seeds is lower than in the 90% fleet restriction policy.

Lastly, Fig. 4.19 depicts the high subsidy extreme policy. For this policy, due to the introduction of high subsidies, electric aircraft become attractive for airlines and conventional aircraft become less attractive. Because of this, there is a decrease in the conventional fleet, as can be seen in the variables in the figure. In this case, as well, the variables show reasonable behavior and do not take any negative values.

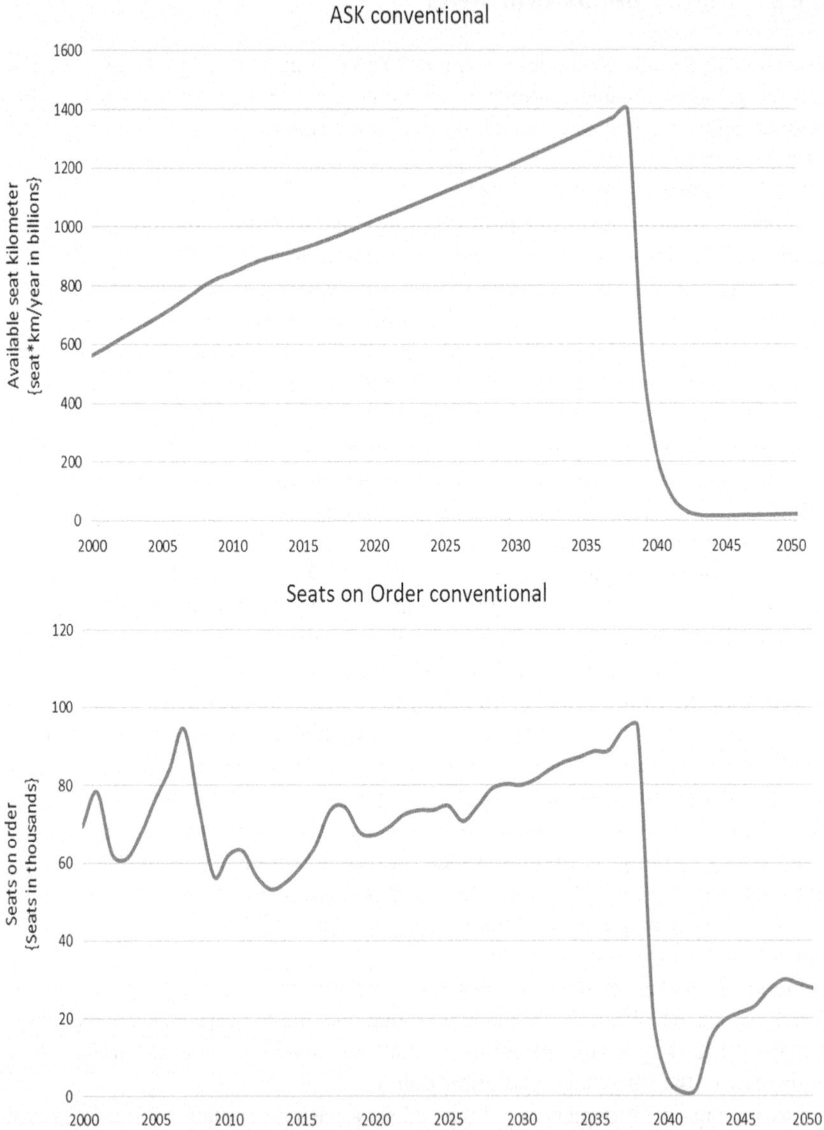

Fig. 4.17 Extreme policy—90% fleet restriction policy

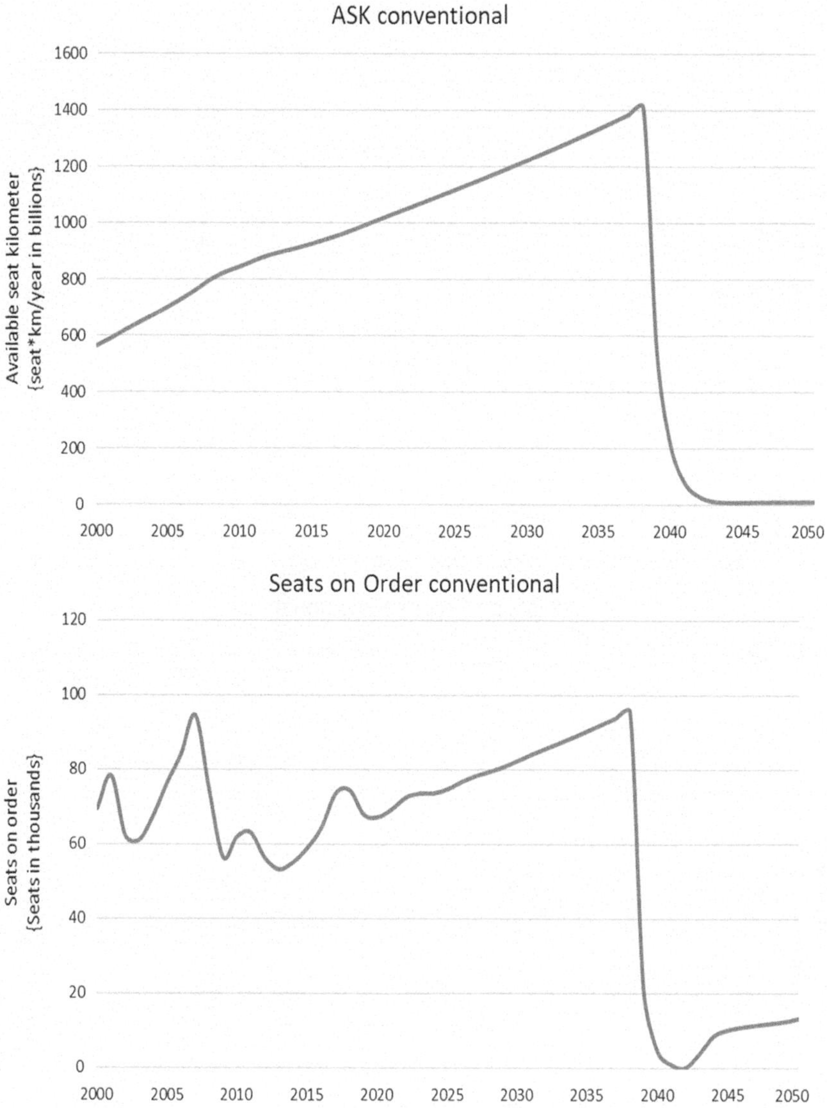

Fig. 4.18 Extreme policy—95% fleet restriction policy

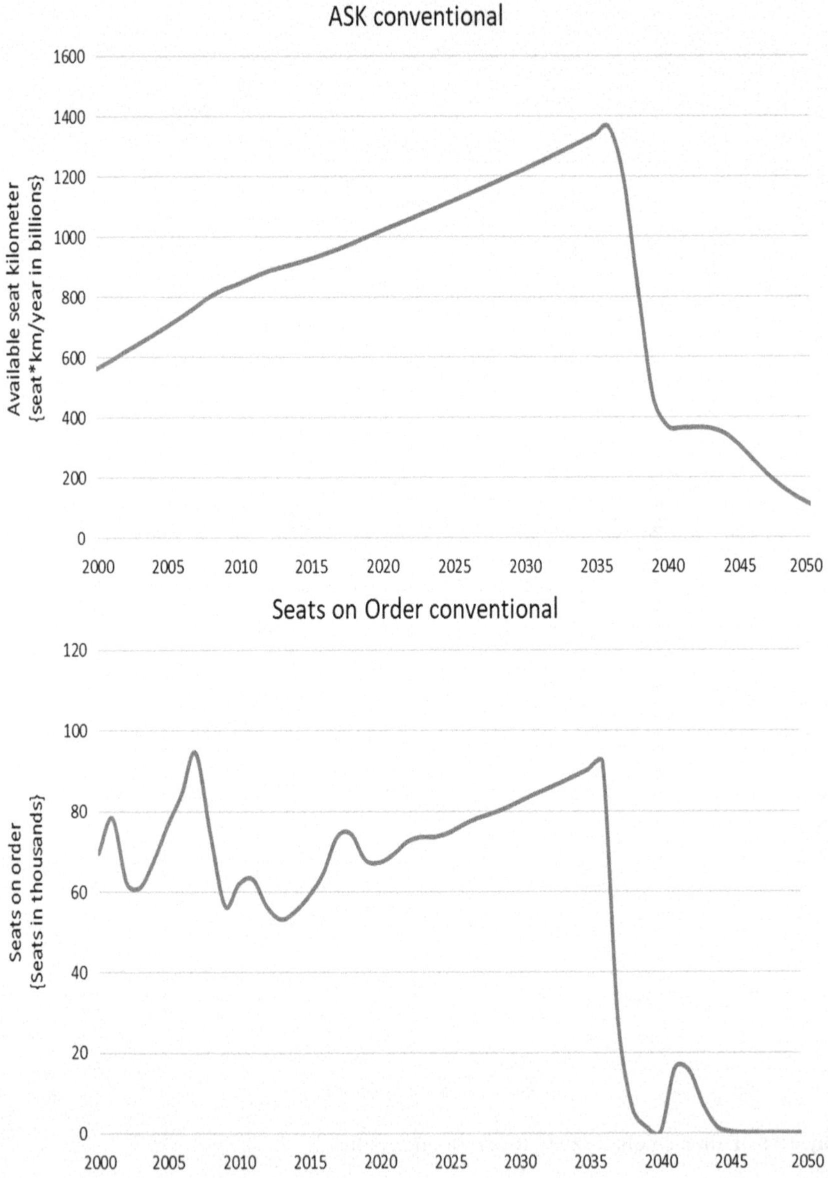

Fig. 4.19 Extreme policy—high subsidy

Results

5

In this section, the results pertaining to the application of the model are presented. The section is divided into two parts and with each having a case study. Each case study begins with an explanation of the main reason for employing the developed model for the particular study. Following that, the goals of the study and the research questions aimed to be addressed through its application are outlined. Subsequently, the scenario generation, policy and strategic settings, along with an explanation of the simulation results, are discussed. Finally, the Monte Carlo sensitivity results are presented to account for the uncertainty of the most important parameters.

5.1 Case Study 1: Introduction of Electric Aircraft in the EU Using Policy Analysis to Reach Flightpath 2050 CO_2 Emissions Goal

In this section, the case study related to the adoption of electric aircraft to help reduce the in-flight CO_2 emissions from the EU is conducted. For this, the model developed in Chap. 4 is utilized. Section 5.1.1, consists of the explanation of the goal of the study along with the research questions. In Sect. 5.1.2, the different simulation scenarios along with simulation results are explained. Lastly, in Sect. 5.1.3, the results of the Monte Carlo sensitivity analysis are discussed. Regarding this case study, the analysis presented below in Sects. 5.1.2 and 5.1.3 is taken from work done in (Talwar et al. 2023).

C. Talwar, *Transition Pathways towards a Low Emission Air Transport System of the Future – A System Level Analysis*, Produktion und Logistik,
https://doi.org/10.1007/978-3-658-46507-0_5

5.1.1 Problem Setting And the Goal of the Case Study

As mentioned in Sect. 2.5, there have been many initiatives taken by regulators and industry professionals to help reduce the harmful impact of emissions from the aviation industry. One such goal is the Advisory Council for Aeronautics Research in Europe's (ACARE) "Flightpath 2050" (European Commission 2017; ACARE 2020). The case study is centered around the emission goals that are set under the Flightpath 2050. One of the goals of this strategy that is relevant to this case study is reducing the European aviation CO_2 emissions by 75% by 2050 compared to the base year 2000. For this particular target, only the emissions pertaining to the direct flight operations, i.e., in-flight emissions were considered (ACARE 2020). In the EU, it has been found that a major portion of the CO_2 emissions come from short-haul flights (Alonso et al. 2014; Grimme and Jung 2018) and narrow body aircraft are most used in this segment. In this regard, it is clear that in order to achieve the emission reduction goals, the short-haul flight segment should be the primary target.

From the discussions in Sect. 2.7, electric aircraft is one of the technologies that can be introduced in the short-haul flight segment. Furthermore, electric aircraft can provide 100% in-flight emission reduction potential. On the other hand, from the discussion in Sect. 2.5 about the different policies that can be introduced in the aviation system it was found that policies like seat tax, jet fuel tax and fleet restriction through quota management are some of the important policies that can force the airlines to move towards sustainability. In the context of this discussion, the aim of this case study is to find out whether the Flightpath 2050 CO_2 emissions goals can be reached using electric aircraft. This is tested in the context of the introduction of jet fuel tax, electricity subsidy, fleet restriction policy and seat tax policy to assess the impact of the policy measures on the adoption rates of electric aircraft. The case study would provide EU policy-makers decision support in reaching the Flightpath 2050 CO_2 emissions goal (Eurocontrol 2018) using narrow body short-haul electric aircraft flights under 1,000 km range.

5.1.2 Parameter Settings and Simulation Results

In this section, the different parameter settings that are required to generate different adoption scenarios are discussed. After discussing the scenarios, the simulation results of the scenarios are shown in the form of a table as well as a diagram.

5.1.2.1 Parameter Settings: Policy Design and Electric Aircraft

The model presented in Chap. 4 is adjusted to the geographical region of the EU by adding the macroeconomic data from the region. Specifically, the model is calibrated with GDP per capita from the EU region along with population to correctly reflect the air travel demand of the region (World Bank 2021). Table 5.1 shows the parameters considered for the development of electric aircraft for this case study. For this, there are two variables namely energy consumption per ASK and initial ownership cost per ASK. To adjust for uncertainty, fast, medium and slow scenarios are developed, where fast is the most optimistic scenario and slow is the least optimistic scenario. For the sake of simplicity and consistency in the analysis, the aircraft pertaining to each scenario are introduced in the year 2035 in the model. The first variable, energy consumption per ASK, is calculated based on the AEA-800 aircraft design developed by (Gnadt et al. 2019) as discussed in Sect. 2.7.

Table 5.1 Parameter values considered for electric aircraft development

Development of electric aircraft			
Variable	Slow	Medium	Fast
Energy consumption per ASK	0.20 kWh/seat*km	0.19 kWh/seat*km	0.17 kWh/seat*km
Initial ownership costs electric per ASK	+50%	+40%	+30%

Table 5.2 Parameter values considered for policy analysis

Policy parameter values			
Policy	Low	Medium	High
Kerosene Tax	0.024 $/l	0.28 $/l	0.53 $/l
Electricity subsidy	0.105 $/kWh	0.124 $/kWh	0.143 $/kWh
Fleet restriction policy	10%	15%	20%
Seat tax	0.013 $/seat*km	0.027 $/seat*km	0.041 $/seat*km

This design assumed that an 800 Wh/kg specific energy battery would be used to power an electric aircraft with 180 passengers and a 926 km range. The amount of total required energy for this aircraft would be 28,800 kWh (Gnadt et al. 2019). Therefore, the energy consumption per seat kilometer for this type of aircraft is given by dividing the energy required by the product of number of seats, i.e., 180, and the number of kilometers, i.e., 926 km. This gives the value of 0.17 kWh/seat*km, which is reflected in the fast development of electric aircraft. Since this number is based on a highly optimistic scenario, the adjustment for uncertainty is done by taking two more scenarios with a 10% and 20% increase in the above value, leading to 0.19 kWh/seat*km and 0.2 kWh/seat*km for the medium and slow scenarios, respectively. The second variable, initial ownership costs per ASK, is based on two aspects: first, the findings by (Ploetner et al. 2013) suggest owning and maintaining the battery system would cost 23% more than the conventional aircraft in terms of costs of ownership. Second, based on the study by (Karpuk and Elham 2021) the cost of maintenance for electric aircraft would be 30% higher than that of conventional aircraft. Adjusting for the share of maintenance cost based on the cost structures and adding up a 30% increase in ownership costs for the electric aircraft is the result, which corresponds to the value in the fast development of the electric aircraft. Similar to the previous variable, the adjustment for uncertainty is done by taking two more scenarios with 10 and 20 percentage points increase in the above value, i.e., 40% and 50% increase for the medium and slow scenarios, respectively. Table 5.2 depicts the parameter values regarding the policy analysis. For each policy, three different cases were considered, which correspond to low, medium and high values. In the case of the kerosene tax, the lower tax rate is based on the current EU maximum kerosene tax limit (European commission, 2021). The highest kerosene tax rate is based on the price recommended by the German federal environment agency. This tax is 180 € per ton of CO_2, when converted in terms of $ per liter of jet fuel the value amounts to 0.53 $/l (Umweltbundesamt 2018, 2019). For the medium tax rate, a middle plausible value was chosen. For the electricity subsidy, the chosen values correspond to a cost coverage of 55%, 65% and 75% of the average electricity price, for the simulation period considered. Taking parallels from the percentage development of the share of the renewable energy sector in the electricity industry in the last two decades, the same values for the fleet restriction policy for the analysis are chosen (Renewable Energy World 2011; EC europa 2020). For the seat tax policy, the values correspond to the cost coverage of 10%, 20% and 30% of the average unit cost to operate conventional aircraft, respectively. Furthermore, uncertainty regarding the development of exogenous factors over time, i.e., EU GDP per capita and EU population, was taken into account.

For this, the values are varied by ± 5% for the scenarios to reflect stronger and weaker growth, analogous to the study by (Kieckhäfer et al. 2018).

5.1.2.2 Simulation Results

Table 5.3 shows the results of the simulation runs with the policy scenarios. First, the policies were tested individually to assess their individual impact on the reduction of aviation emissions. It can be observed in Table 5.3 that the individual policy able to reach the Flightpath 2050 CO_2 emissions goals are the high electricity subsidy and fleet restriction policy. On the other hand, high kerosene tax was not able to reduce emissions below 75% of the year 2000 value. Furthermore, the low and medium kerosene tax and electricity subsidy had very low or no effect on the reduction of CO_2 emissions. Therefore, they are not included in the analysis. Next, a few combinations of policies were tested to check for their efficacy in reducing CO_2 emissions. It can be observed in Table 5.3 that most of the policy combinations achieve the Flightpath 2050 CO_2 emissions goals. The extent to which the emissions are reduced and the electric fleet is increased depends on the magnitude of the policy parameter. For instance, the application of a combination of a low fleet restriction and a seat tax results in 56% of the CO_2 emissions value at the end of the simulation period (2050) as compared to the base year 2000. Furthermore, it can be observed that the overall percentage of the electric fleet operated at the end of the simulation period is 13%. Figure 5.1 depicts the development of emissions and electric fleet in the model for the low fleet restriction and seat tax policy. In this figure, the gray area represents the time when the policy was implemented in the model, i.e., 2038. In the case of the high subsidy and low fleet restriction policy scenario, the CO_2 emissions value is 37% and the share -of the electric fleet is 43% of the overall fleet. Figure 5.2 shows the development of emissions and electric fleet over time in the model for this particular policy. The highest reduction in CO_2 emissions using a combination of policies was achieved in the case of a high subsidy and high kerosene tax policy. Specifically, in this scenario, the CO_2 emission value was 7% and the share of the electric fleet was 89%. Apart from the combination of policies, a pro-electric scenario was tested where all the policy parameters are set at the highest value to promote electric fleet share. This particular scenario implicitly represents a situation where the policy-makers highly encourage the electric fleet. For this scenario, it can be observed in Table 5.3 that the CO_2 emissions value is 8% and the share of the electric fleet is 89%. Furthermore, the best and the worst case scenarios were also tested. In the best case scenario, the exogenous and policy parameters are set in favor of a higher electric fleet adoption, whereas in the worst case scenario, these parameters are set in favor of

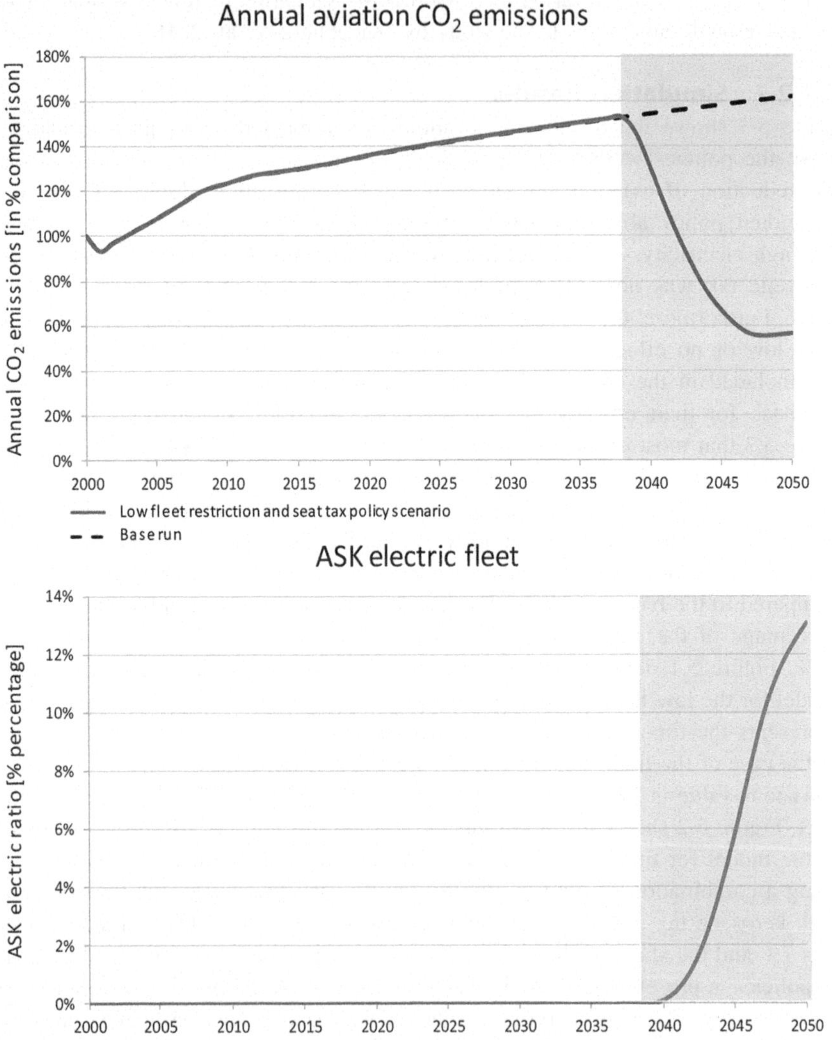

Fig. 5.1 Result overview of low fleet restriction and seat tax policy

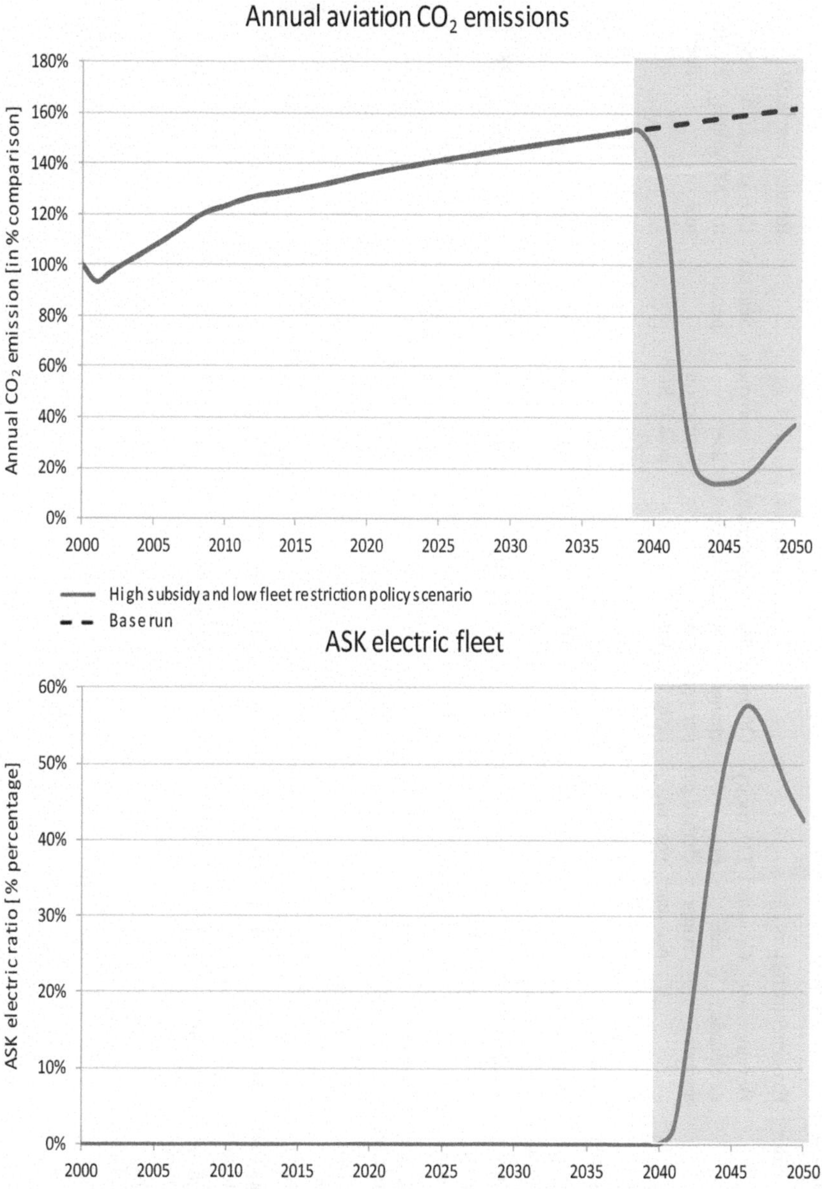

Fig. 5.2 Result overview of high subsidy and low fleet restriction policy

Table 5.3 Overview of simulation results

Scenarios	Exogeneous factors				Policy measures				Results year 2050	
	Development of electric aircraft	Growth rate change per year in%	Electricity price change per year in%	Oil price change per year in%	Jet fuel tax rate	Electricity subsidy	Government fleet restriction policy 2038 in%	Seat tax penalty	CO_2 compared to 2000 values in%	ASK electric fleet in%
Base run	Medium	0	0	0	–	–	0	–	162	0
High jet fuel tax	Medium	0	0	0	High	–	0	–	160	0
High electricity subsidy	Medium	0	0	0	–	High	0	–	74	28
Low fleet restriction	Medium	0	0	0	–	–	10	–	57	13
Medium fleet restriction	Medium	0	0	0	–	–	15	–	39	23
High fleet restriction	Medium	0	0	0	–	–	20	–	29	34
High jet fuel tax and subsidy	Medium	0	0	0	High	High	0	–	7	89

(continued)

Table 5.3 (continued)

Scenarios	Exogeneous factors				Policy measures				Results year 2050	
	Development of electric aircraft	Growth rate change per year in%	Electricity price change per year in%	Oil price change per year in%	Jet fuel tax rate	Electricity subsidy	Government fleet restriction policy 2038 in%	Seat tax penalty	CO_2 compared to 2000 values in%	ASK electric fleet in%
Low fleet restriction and low seat tax	Medium	0	0	0	–	–	10	Low	56	13
Low fleet restriction and high seat tax	Medium	0	0	0	–	–	10	High	35	45
High fleet restriction and low seat tax	Medium	0	0	0	–	–	20	Low	29	34
High fleet restriction and seat tax	Medium	0	0	0	–	–	20	High	27	51

(continued)

Table 5.3 (continued)

Scenarios	Exogeneous factors				Policy measures				Results year 2050	
	Development of electric aircraft	Growth rate change per year in%	Electricity price change per year in%	Oil price change per year in%	Jet fuel tax rate	Electricity subsidy	Government fleet restriction policy 2038 in%	Seat tax penalty	CO_2 compared to 2000 values in%	ASK electric fleet in%
High subsidy and low fleet restriction	Medium	0	0	0	–	High	10	–	37	43
High subsidy and fleet restriction	Medium	0	0	0	–	High	20	–	25	56
Pro electric	Fast	0	0	0	High	High	20	–	8	89
Best case scenario	Fast	−5	−10	10	High	High	20	High	11	82
Worst case scenario	Slow	5	10	−10	–	–	0	–	192	0

Table 5.4 Overview of remainder of scenarios case study 1

Scenarios	Exogenous factors				Policy measures				Results year 2050	
	Development of electric aircraft	Growth rate change per year in%	Electricity price change per year in%	Oil price change per year in%	Jet fuel tax rate	Electricity subsidy	Government fleet restriction policy 2038 in%	Seat tax penalty	CO_2 compared to 2000 values in%	ASK electric fleet in%
Medium kerosene and seat tax	Fast	0	0	0	Medium	–	0	Medium	32	46
Medium jet fuel tax and High fleet restriction	Fast	0	0	0	Medium	–	20	–	104	99
Medium jet fuel, seat tax and fleet restriction	Fast	0	0	0	Medium	–	15	Medium	6	99
Medium jet fuel tax and electricity subsidy and high seat tax	Fast	0	0	0	Medium	Medium	0	High	6	90

(continued)

Table 5.4 (continued)

Scenarios	Exogenous factors				Policy measures				Results year 2050	
	Development of electric aircraft	Growth rate change per year in%	Electricity price change per year in%	Oil price change per year in%	Jet fuel tax rate	Electricity subsidy	Government fleet restriction policy 2038 in%	Seat tax penalty	CO_2 compared to 2000 values in%	ASK electric fleet in%
High jet fuel tax, fleet restriction and seat tax	Medium	0	0	0	High	–	20	High	23	99
High jet fuel tax, low electricity subsidy and medium seat tax	Medium	0	0	0	High	Low	0	Medium	7	90
Low jet fuel tax and seat tax	Medium	0	0	0	Low	–	0	Low	161	0

(continued)

Table 5.4 (continued)

Scenarios	Exogenous factors				Policy measures				Results year 2050	
	Development of electric aircraft	Growth rate change per year in%	Electricity price change per year in%	Oil price change per year in%	Jet fuel tax rate	Electricity subsidy	Government fleet restriction policy 2038 in%	Seat tax penalty	CO_2 compared to 2000 values in%	ASK electric fleet in%
Low jet fuel tax, medium electricity subsidy and seat tax	Medium	0	0	0	Low	Medium	0	Medium	13	80
Medium jet fuel tax and fleet restriction	Slow	0	0	0	Medium	–	15	–	104	99

(continued)

Table 5.4 (continued)

Scenarios	Exogenous factors				Policy measures				Results year 2050	
	Development of electric aircraft	Growth rate change per year in%	Electricity price change per year in%	Oil price change per year in%	Jet fuel tax rate	Electricity subsidy	Government fleet restriction policy 2038 in%	Seat tax penalty	CO_2 compared to 2000 values in%	ASK electric fleet in%
Medium jet fuel tax, electricity subsidy and seat tax and low fleet restriction	Slow	0	0	0	Medium	Medium	10	Medium	10	91
Medium jet fuel tax, low electricity subsidy and high seat tax	Slow	0	0	0	Medium	Low	0	High	27	60

(continued)

Table 5.4 (continued)

Scenarios	Exogenous factors				Policy measures				Results year 2050	
	Development of electric aircraft	Growth rate change per year in%	Electricity price change per year in%	Oil price change per year in%	Jet fuel tax rate	Electricity subsidy	Government fleet restriction policy 2038 in%	Seat tax penalty	CO_2 compared to 2000 values in%	ASK electric fleet in%
Medium jet fuel tax, electricity subsidy, fleet restriction and seat tax	Slow	0	0	0	Medium	Medium	15	Medium	8	94
High jet fuel tax and low electricity subsidy and fleet restriction policy and seat tax	Slow	0	0	0	High	Low	10	Low	12	86

(continued)

Table 5.4 (continued)

Scenarios	Exogenous factors				Policy measures				Results year 2050	
	Development of electric aircraft	Growth rate change per year in%	Electricity price change per year in%	Oil price change per year in%	Jet fuel tax rate	Electricity subsidy	Government fleet restriction policy 2038 in%	Seat tax penalty	CO_2 compared to 2000 values in%	ASK electric fleet in%
High jet fuel tax, medium electricity subsidy and low seat tax	Slow	0	0	0	High	Medium	0	Low	5	90
Low jet fuel tax and fleet restriction, medium seat tax	Slow	0	0	0	Low	–	10	Medium	105	98
Low jet fuel tax, electricity subsidy and seat tax	Slow	0	0	0	Low	Low	0	Low	31	42

higher conventional fleet. Table 5.3 shows that the CO_2 emissions in the best case scenario are 11% compared to the 2000 value along with a total 82% share of the electric fleet. On the other hand, in the worst case scenario, the CO_2 emissions levels are at 192% with a 0% electric fleet. All other combination scenarios not in Table 5.3 are presented in Table 5.4. To make the results more robust, different introduction years of the major policies fleet restriction, seat tax and electricity subsidy between the years 2035 and 2040 were also tested.

One of the main reasons to select this particular time frame to test for different years testing is that if the electric aircraft are introduced in the year 2035 then it does not make meaningful sense for the policy to be introduced after 2040. It was found that the results for this analysis were not too different from the results for 2038 shown in Table 5.3. One reason for this could be that the airline industry is dominated by delays, where the industry exhibits long cycles of replenishment of fleet (Liehr et al. 2001; Sgouridis et al. 2011; Pierson and Sterman 2013). Hence, a span of five years might not be able to show huge differences in the results.

5.1.3 Sensitivity Analysis

In system dynamics, to study the robustness of the model, numerical sensitivity analysis is often used because it is considered the most important for addressing future scenarios (Sterman 2000). For this case study, Monte Carlo simulation is used to determine the probability of the occurrence of different simulation results that are caused by different parameter values of the exogenous factors. In 500 simulation runs, the parameters of the exogenous factors are varied in an interval. Each simulation run is then saved and provides a confidence interval. The Monte Carlo simulations are carried out in *Vensim DSS version 8.0.8*. Table 5.5 shows the interval bounds of exogenous variables that have been used to conduct

Table 5.5 Interval bounds of variables used for Monte Carlo sensitivity simulation.

Variable	Distribution	Low	High
Energy consumption per ASK	Random uniform	0.17	0.2
Initial ownership costs per ASK	Random uniform	+30%	+50%
Growth factor% (GDP per capita, population)	Random uniform	−5%	+5%
Jet fuel price growth%	Random uniform	−10%	+10%
Electricity price growth%	Random uniform	−10%	+10%

the sensitivity analysis. These particular variables are chosen for the sensitivity analysis as they are the factors that are arguably highly uncertain and likely to be more influential (Sterman 2000). Figure 5.3 shows the results of the multivariate sensitivity analysis for the base case scenario. The results suggest that given the distribution and interval for the exogenous variables in, there is a 100% chance that the CO_2 emissions at the end of the simulation period (2050) lie between 103 million tonnes to 146 million tonnes if no policies are implemented. This translates to 136% to 192% emission levels compared to the base year 2000. In the univariate sensitivity analysis for the base case scenario, it was observed that the model is more sensitive to the variable growth factor and not to the variables related to the electric fleet. One of the reasons for this is that in the base case scenario, electric aircraft are extremely expensive to operate as compared to conventional aircraft. Due to this, any changes in the parameters related to the electric fleet do not have any effect on the adoption of electric aircraft.

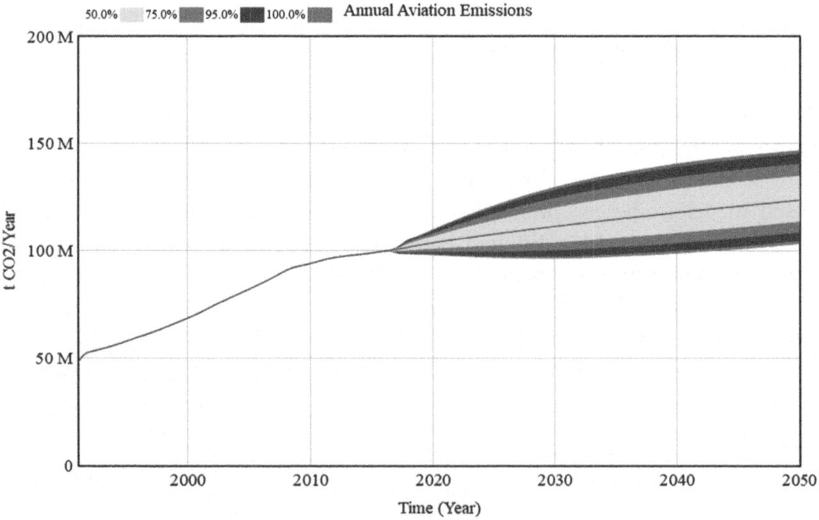

Fig. 5.3 Multivariate sensitivity results for base run

In the case of the variable growth factor, any change directly alters the air travel demand and so the airlines have to adjust their fleet operations. The different levels of conventional fleet operation due to adjustments ultimately affect the

CO_2 emissions levels, thereby causing sensitivity. Figure 5.4 shows the multivariate sensitivity analysis for low seat tax and fleet restriction policy. In this case, the Monte Carlo sensitivity analysis predicts a 100% confidence interval between 37 million tonnes and 50 million tonnes of CO_2 emissions. This translates to an emission value between 49% and 66% of the base year 2000 levels. Furthermore, the univariate sensitivity analysis for this policy showed a high sensitivity only for the growth factor variable. It can be argued that this is because the policy is stringent in nature and focused more on reducing conventional fleet than promoting the electric fleet through subsidy. Hereby, the airlines are forced to shift towards an electric fleet regardless of the changes in the development costs for electric aircraft and electricity prices. Thus, the model only showed sensitivity for variables like EU GDP per capita and EU population but not to the parameters related to electric aircraft.

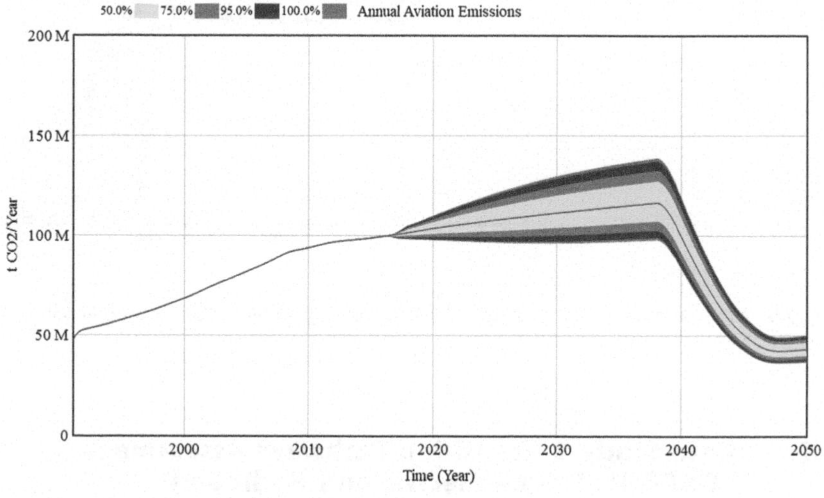

Fig. 5.4 Multivariate sensitivity results for low seat tax and fleet restriction policy

The multivariate sensitivity analysis for high subsidy and low fleet restriction policy is depicted in Fig. 5.5. The confidence interval here is between 2 and 50 million tonnes, i.e., between 3% and 66% of the annual aviation CO_2 emissions of the year 2000 levels. In the univariate analysis, all the considered parameters showed sensitivity in the model. A possible reason for a larger confidence interval and sensitivity to the mentioned factors could be the less stringent nature of the policy. Unlike low fleet restriction and seat tax policy, this particular policy is aimed at encouraging the electric fleet by providing subsidies. Therefore, any changes in the ownership variables, energy consumption efficiency of electric aircraft, electricity price and/or jet fuel price can cause volatility in the system and can further lead to volatility in the adoption of electric aircraft, which potentially causes sensitivities in the confidence interval of CO_2 emissions.

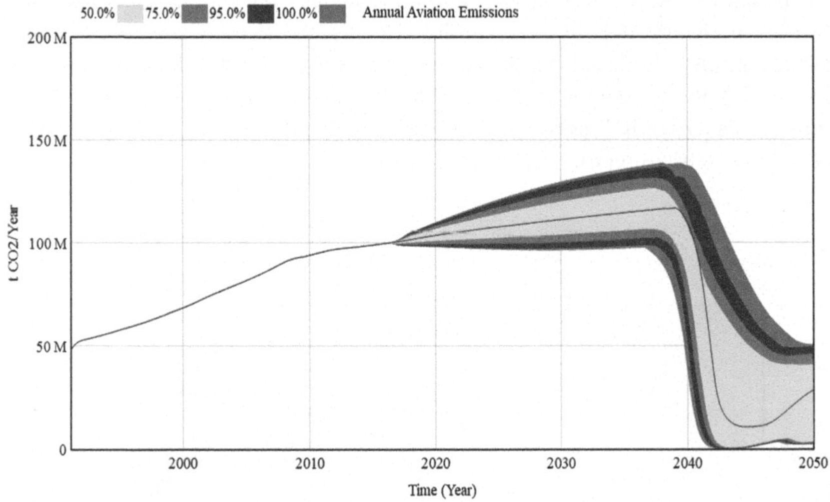

Fig. 5.5 Multivariate sensitivity analysis for high subsidy and low fleet restriction policy

5.2 Case Study 2: Transition Pathways Assessment of SAF, Hydrogen-Electric and Hydrogen Combustion in the ATS Model

In this section, a case study examining the adoption of SAF, hydrogen-electric, and hydrogen combustion aircraft across various flight range segments is undertaken. For this, the model developed in Chap. 4 is utilized. Section 5.2.1 elucidates the study's objectives and research questions. Following that, Sect. 5.2.2 delineates the different simulation scenarios and their corresponding results. Finally, Sect. 5.2.3 delves into the findings of the Monte Carlo sensitivity analysis. Regarding this case study, the analysis presented below in Sects. 5.2.2 and 5.2.3 is taken from work done in (Joormann et al. 2024).

5.2.1 Problem Setting and the Goal of the Case Study

In Sect. 2.5, IATA's net zero carbon strategy was discussed and the industry regulators aim for the aviation system to be net zero by 2050. IATA believes that 65% of this reduction might come from sustainable aviation fuel (SAFs) and 13% from the introduction of new propulsion technologies. The case study is aimed at unraveling the strategies through the adoption of new technologies and scaling of fuel production that would be required to achieve the net zero emissions goal by 2050. In this context, in Sect. 2.6 and Sect. 2.7, discussions were done regarding the potential of SAFs and hydrogen technology in helping to achieve the climate goals set by IATA for the industry. According to reports and expert opinions, it is projected that by 2030, 5% of the global jet fuel demand could potentially be replaced by SAF (World Economic Forum 2020). By 2050, estimates suggest that 330 to 445 million tons of SAF could be produced, primarily through hydrogen-based power-to-liquid (PtL) production pathways, to substitute for 65% of the global jet fuel demand (ATAG 2020b; World Economic Forum 2020). Concerning hydrogen propulsion-based aircraft, reports and expert opinions indicate that between 2035 and 2040, approximately 8% to 16% of the total global fleet could transition to hydrogen aircraft. Additionally, 40% to 60% of the fleet might be replaced by hydrogen technology by 2050. This transition may necessitate 5 to 130 million tons of hydrogen per year for the aviation system from 2035 to 2050 (Fuel Cells and Hydrogen Joint Undertaking 2020). Furthermore, research indicates that SAF can be utilized across short, medium, and long-haul flight segments, while hydrogen fuel-cell technology may find application in short-haul flights and hydrogen combustion technology in medium and long-haul flights (Fuel Cells and Hydrogen Joint Undertaking 2020).

Considering these findings, it is clear that there is future potential for the two technologies but there has been limited comprehensive systems analysis of their combined impact on the system. Moreover, predicting the various transition pathways toward a more sustainable future for air transport is challenging due to uncertainties such as delays in SAF production scalability, hydrogen availability as a fuel, and the accessibility of green electricity for green hydrogen production in the aviation sector. Also, it is unclear how these technologies would be adopted in different range segments to help achieve the net zero goal by 2050. Therefore, the primary objective of this case study is to analyze the various diffusion strategies of new aircraft propulsion technologies, namely SAF, hydrogen electric, and hydrogen combustion in three different range segments namely short, medium and long-haul. To do this, multiple scenarios are generated to examine diffusion rates considering factors such as delays in hydrogen aircraft and fuel manufacturing,

SAF manufacturing delays, macroeconomic fluctuations, shifts in flight segment distribution, scale effects resulting from production increases, and variations in green electricity availability. By evaluating the outcomes of these scenarios, the case study would provide SAF, hydrogen and electricity requirements for the different transition strategies.

5.2.2 Parameter Settings and Simulation Results

This section examines the different parameter settings to develop various adoption scenarios. Following the discussion of these scenarios, the simulation outcomes are presented both in tabular format and through diagrams.

5.2.2.1 Parameter Settings: Hydrogen Production, Green Electricity, SAF and Flight Segment Share

The extended model shown in Chap. 4 is fitted with the relevant data from multiple sources to test for the different transition scenarios for the future ATS. To correctly reflect the global air travel demand, the extended model is calibrated with global GDP per capita and population levels (World bank 2023a, b). Furthermore, to realistically represent the ticket prices, the model has also been fitted with the consumer price inflation data and price of conventional jet fuel taken over from (Kieckhäfer et al. 2018). Table 5.6 shows the different initial level parameter settings for the simulation runs. Since the model has a large number of input parameters, only the most important ones are discussed and showcased in this section. Based on the discussion in Sect. 2.6, it is found that SAFs can be used in place of conventional jet fuel and do not require any structural changes to the aircraft or to airport refueling infrastructure. Various bioconversion technologies exist for the production of SAF, with some being well established methods, while others are relatively new and require further advancements. Irrespective of the manufacturing process, SAF (when blended with traditional jet fuel) must meet the stringent requirements set forth by certifying organization i.e., the American Society for Testing and Materials (ASTM), to be considered as a drop-in fuel (Shahriar and Khanal 2022). The major SAF production pathways include Fischer Tropsch synthesis (FT-SPK), hydroprocessed fatty acid esters and fatty acids (HEFA), alcohol to jet (ATJ) and power to liquid (PtL) (Zhang et al. 2020; Tanzil et al. 2021; IATA 2022b; Cabrera and Melo de Sousa 2022; Hsu et al. 2022). For this case study, the share of each type of SAF that can be expected by 2050 is included in the model as an exogenous parameter based on the findings of reports and expert opinions (ATAG 2020b; World Economic Forum 2020; ICCT

2021; Eurocontrol 2022b). SAFs are introduced in 2025 in the model, the initial amount of SAF capacity available in the system is taken to be 7.5 billion liters (ATAG 2020b; IATA 2022b). Since hydrogen is required to produce the PtL type of SAF, the initial parameter for the hydrogen requirement is taken to be 0.439 kg of hydrogen per liter of PtL type SAF (Fly zero 2022). Furthermore, the electricity required to produce green hydrogen is set to 55 kWh per kg of hydrogen produced based on recent literature findings (Kakoulaki et al. 2021). For hydrogen technology, from the current level of research, it is found that there are two ways to use hydrogen in aircraft. First, through thermal conversion in H_2 combustion engines and second, through electro chemical conversion in fuel-cells. The new propulsion concepts would also require the integration of hydrogen storage on board. Due to this reason to make these aircraft a reality, new aircraft designs would be necessary. In light of different energy, densities and propulsion system capabilities, it is believed that hydrogen combustion can be potentially used for medium and long range aircraft, whereas hydrogen fuel-cells might be applicable for short hall flight range segment (Fuel Cells and Hydrogen Joint Undertaking 2020; Hoelzen et al. 2022b).

For this reason, the short range fuel-cell aircraft is introduced in the year 2035 in the system, whereas medium and long range hydrogen combustion aircraft are introduced in 2040 and 2045 (Fuel Cells and Hydrogen Joint Undertaking 2020). The data related to fleet segregation into three different ranges is taken over from the study conducted by (Hoelzen et al. 2022a). Regarding hydrogen availability, the initial level of hydrogen availability from 2030 onwards is taken from (Fuel Cells and Hydrogen Joint Undertaking 2020). Correspondingly, the expected share of green hydrogen and green electricity availability is taken from the net zero scenarios conducted by the IEA in their hydrogen review report (IEA 2021b). The improvement in the efficiency of hydrogen technology in aircraft is assumed to be similar to jet fuel technology over the years after being introduced in the system in 2035 (Joormann et al. 2024). Lastly, different adoption scenarios of each technology are simulated in the model. These scenarios are plugged into the model as an input variable and the value of these inputs are decided on the basis of the different expected future developments of SAF and hydrogen technology in aircraft from research reports and existing literature. This sort of open loop simulation testing helps in assessing the interaction between the model components and effect of delays in the system when a certain technology is adopted in the system.

Table 5.6 Exogenous parameters

Parameter setting for exogenous variables				
Parameter	Brief description	Value for simulation	Introduction year	Source
Introduction of H_2 aircraft	Introduction year for H_2 aircraft in different flight segments	N.A.	2035: Short range fuel-cell 2040: Medium range H_2 combustion 2045: Long range H_2 combustion	(Fuel Cells and Hydrogen Joint Undertaking 2020; Hoelzen et al. 2022a)
Hydrogen availability	Initial level of H_2 availability for aviation usage	5 Million tons	2035	(Fuel Cells and Hydrogen Joint Undertaking 2020; IEA 2021b)
Fleet segregation	Data used for the segregation of fleet into different ranges	63%—Short-haul 14%—Medium-haul 16%—Long-haul	N.A.	(Hoelzen et al. 2022a)
Green H_2 share	Share of green H_2 in the overall H_2 capacity	30% of total H_2 availability	2030–2035	(IEA 2021b)
Electricity required to produce green H_2	Amount of electricity in KWh required to produce 1 kg of H_2	55 kWh/kg	N.A.	(Kakoulaki et al. 2021)
Green electricity availability	Amount of initial green electricity that is available to produce green H_2	1.85 TWh	By 2030	(IEA 2021b)

(continued)

Table 5.6 (continued)

Parameter setting for exogenous variables				
Parameter	Brief description	Value for simulation	Introduction year	Source
Share of different types of SAF	Percentage share of different types of SAF in the system	**Until 2030:** HEFA: 95% FT synthesis: 0.5% ATJ: 4% PtL: 0.5% **Until 2050:** HEFA: 14% FT synthesis: 10% ATJ: 14% PtL: 62%	N.A.	(ATAG 2020b; ICCT 2021; Eurocontrol 2022b)
Initial amount of SAF	Amount of SAF available in the system at the starting simulation point	7.5 billion liters	2025	(ATAG 2020b; Eurocontrol 2022b)
Hydrogen required per liter of PtL SAF	Amount of hydrogen required to produce one liter of PtL SAF based on current technology	0.439 Kg/liter	N.A.	(Fly zero 2022)
Improvement rate	Improvement in the fuel consumption efficiency of hydrogen and jet fuel technology in aircraft	Hydrogen: 1.5%	2035	(Roland Berger 2020b) and assumption

5.2.2.2 Simulation Results

Base simulation run

The base scenario depicts the business as usual case for the air transport industry. It does not contain the introduction of new propulsion technologies and showcases the simulation results where growth in the industry happens similar to past years. Figure 5.6 (a) displays the total available seat kilometers (ASK) in the system as a percentage comparison to the year 2025 and separation of the ASK by the flight segments. Figure 5.6 (b) shows the share of available seat kilometers of the different flight segments.

Figure 5.6 (c) depicts the fuel consumption by flight segment it can be seen that the ASK is 185% of the 2025 levels and most of the traffic comes from the short range segment. The overall fuel consumption is 566 billion liters, out of which 390 billion liters, 51 billion liters and 80 billion liters are used for short, medium and long-haul flight segments, respectively.

Simulation Results

As discussed in Sect. 5.2.1, the scenarios chosen for the simulation are decided based on the expected percentage development of the new potential technologies in the system by 2050. Table 5.7 shows the simulation results of the different scenarios conducted in this study. The scenarios for hydrogen technology vary between 20–60% share of the fleet in the system, while, the scenario for SAF varies between 20–65% of alternative fuel in the system. A combination scenario where both technologies co-exist in the system is also tested in the model.

The assessment variables shown in Table 5.7, i.e., hydrogen capacity by 2050, CO_2 emissions from H_2 production, CO_2-eq emissions from H_2 production 2050 and green electricity requirements for green H_2 are chosen to analyze the different requirements and emissions of the simulation runs. The simulation results in Table 5.7 show that 20–60% hydrogen fleet share would require 33–95 billion kg of hydrogen fuel. Producing this amount of hydrogen results in 20–60 million tons of CO_2 emissions in 2050 according to current capabilities. The CO_2-equivalent emissions for the same are between 63–183 million tons. According to scenarios by the IEA hydrogen review report, if the share of green hydrogen in the total amount of produced hydrogen for aviation is 30%, then 290–798 TWh of green electricity would be required for production by 2050. The results for the 20–65% SAF scenarios show that 26–82 billion kg of hydrogen might be required to produce the PtL type of SAF in the system. Note that within these scenarios, the PtL share of SAF types is set to be 62% by 2050. In this case, it would result in 16–51 million tons of CO_2 and 51–158 million tons of CO_2-equivalent emissions. In terms of electricity requirements, 305–961 TWh of electricity might be necessary to meet the demands by

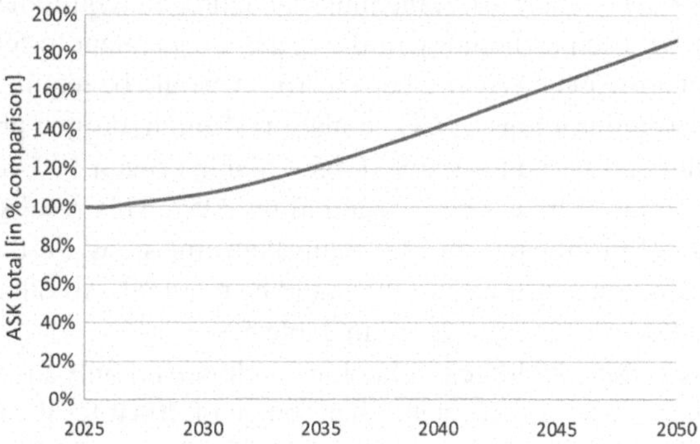

(a) Available seat kilometer base run

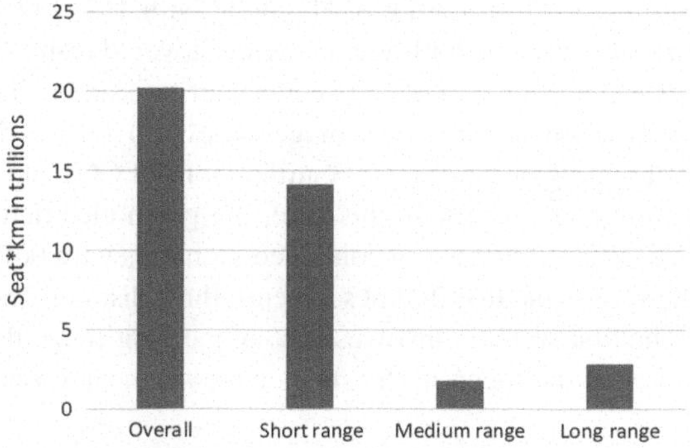

(b) Available seat kilometer 2050 by flight segment

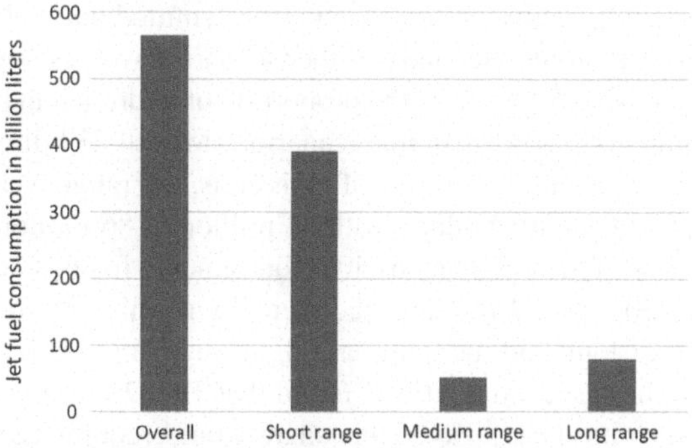

(c) Jet fuel consumption 2050 by flight segment

Fig. 5.6 Base run from 2025 until 2050

2050. In the combination scenarios, the minimal hydrogen requirement is observed for a 20% share for each technology in the system, equivalent to 59 billion kg by 2050. In this scenario, 594 TWh of green electricity would be required by 2050 and would result in 36 million tons of CO_2 emissions from hydrogen production. The maximal amount is observed for the SAF 65% and hydrogen 10% scenario where almost 100 billion kg of hydrogen is required by 2050. This results in 61 million tons of CO_2 and 191 million tons of CO_2-equivalent emissions. Furthermore, in this scenario, 1,121 TWh of green electricity would be required. All other combination scenarios not in Table 5.7 are presented in Table 5.8.

In the scenarios where hydrogen technology is chosen for only short and medium-haul flight segments with a percentage share range of 20–40%, it can be observed that hydrogen and electricity requirements are much lower than in the overall and combination simulation scenarios. Specifically, in the 20–40% share scenarios, the system could require 13–23 billion kg of H_2 along with 157–244 TWh of green electricity. In terms of emissions for these scenarios, it would result in 9–15 million tons of CO_2 and 26–45 million tons of CO_2-equivalent emissions. The scenarios for SAF in the long range segment with a percentage share of 80–90% of the fleet require 11–13 billion kg of hydrogen, emitting 7–8 million tons of CO_2 and 23–25 million tons of CO_2-equivalent emissions. In this case, the green electricity requirement was 131–150 TWh. When considering combined scenarios with both technologies existing within the system but in different segments, three distinct types of scenarios were simulated. The first scenario involves 90% of the long-range fleet using SAF, while the proportion of hydrogen in the short-range segment varies from 20% to 60%.

In this scenario, the hydrogen requirement was found to be 24–46 billion kg along with 15–28 million tons of CO_2 emissions to produce hydrogen. Furthermore, 46–88 million tons of CO_2-equivalent emissions were emitted and 269–455 TWh of green electricity was required. Second, another scenario involves 80% operation of the long-range fleet using SAF, while the proportion of hydrogen in the short-range segment varies from 20% to 60%. In this scenario, a total of 23 billion to 45 billion kg of hydrogen were required. In terms of emissions, the range was 14 million to 27 million tons of CO_2 emitted, along with 44 million to 86 million tons of CO_2-equivalent emissions. The green electricity requirements for this set of scenarios were found to be 250–436 TWh. Third, the scenario in which no short-haul flights exist and 80–90% of long and medium-haul flight segments are replaced by SAF was simulated. In these scenarios, it was found that 19–22 billion kg of hydrogen was required, and to produce 30% of this share would require 227–258 TWh of green electricity.

Table 5.7 Overview of simulation results

Scenarios	Flight segment	Hydrogen capacity 2050 (Kg per year in billion)	CO_2 emissions from H_2 production 2050 (million tonnes)	CO_2 eq emissions from H_2 production 2050 (million tonnes)	Green electricity requirements for green H_2 (TWh per year)
Base run		0	0	0	0
Hydrogen 20%—Overall	Overall	33	20	63	290
Hydrogen 30%—Overall	Overall	48	30	93	413
Hydrogen 40%—Overall	Overall	64	40	124	546
Hydrogen 60%—Overall	Overall	95	60	183	798
SAF 20%—Overall	Overall	26	16	51	305
SAF 30%—Overall	Overall	39	24	75	450
SAF 40%—Overall	Overall	51	32	99	596
SAF 65%—Overall	Overall	82	51	158	961
SAF 20%—Overall and Hydrogen 20%—Overall	Overall	59	36	114	594
SAF 30%—Overall and Hydrogen 30%—Overall	Overall	86	54	168	864
SAF 40%—Overall and Hydrogen 20%—Overall	Overall	84	52	162	886
SAF 65%—Overall and Hydrogen 10%—Overall	Overall	99	61	191	1,121
Hydrogen Short and Medium-haul 20%	Short-haul	11	7	21	119

(continued)

Table 5.7 (continued)

Scenarios	Flight segment	Hydrogen capacity 2050 (Kg per year in billion)	CO_2 emissions from H_2 production 2050 (million tonnes)	CO_2 eq emissions from H_2 production 2050 (million tonnes)	Green electricity requirements for green H_2 (TWh per year)
	Medium-haul	2	2	5	38
Hydrogen Short-haul 30% and Medium-haul 10%	Short-haul	16	10	31	165
	Medium-haul	2	1	3	33
Hydrogen short-haul 30% and medium-haul 30%	Short-haul	16	10	31	165
	Medium-haul	3	2	6	43
Hydrogen short-haul 40% and medium-haul 10%	Short-haul	22	14	42	211
	Medium-haul	1	1	3	33
SAF long-haul 90%	Long-haul	13	8	25	150
SAF long-haul 80%	Long-haul	11	7	23	131
SAF long-haul 90% and Hydrogen short-haul 20%	Short-haul	11	7	21	119
	Long-haul	13	8	25	150
SAF long-haul 90% and Hydrogen short-haul 30%	Short-haul	16	10	31	165
	Long-haul	13	8	25	150
SAF long-haul 90% and Hydrogen short-haul 40%	Short-haul	22	14	42	211

(continued)

Table 5.7 (continued)

Scenarios	Flight segment	Hydrogen capacity 2050 (Kg per year in billion)	CO_2 emissions from H_2 production 2050 (million tonnes)	CO_2 eq emissions from H_2 production 2050 (million tonnes)	Green electricity requirements for green H_2 (TWh per year)
	Long-haul	13	8	25	150
SAF long-haul 90% and Hydrogen short-haul 60%	Short-haul	33	20	63	305
	Long-haul	13	8	25	150
SAF long-haul 80% and Hydrogen short-haul 20%	Short-haul	11	7	21	119
	Long-haul	12	7	23	131
SAF long-haul 80% and Hydrogen short-haul 30%	Short-haul	16	10	31	165
	Long-haul	12	7	23	131
SAF long-haul 80% and Hydrogen short-haul 40%	Short-haul	22	14	42	211
	Long-haul	12	7	23	131
SAF long-haul 80% and Hydrogen short-haul 60%	Short-haul	33	20	63	305
	Long-haul	12	7	23	131
SAF long and medium-haul 90%	Medium-haul	9	6	18	108
	Long-haul	13	8	25	150
SAF long and medium-haul 80%	Medium-haul	8	5	16	96
	Long-haul	11	7	23	131

Table 5.8 Overview of remainder of scenarios case study 2

Scenarios	Assessment variables			
	Hydrogen capacity 2050 (Kg per year in billion)	CO_2 emissions from H_2 production 2050 (Million tonnes)	CO_2-eq emissions from H_2 production 2050 (Million tonnes)	Green electricity requirements for green H_2 (TWh per year)
Base run	0	0	0	0
Hydrogen 10%	17	11	33	161
Hydrogen 50%	83	52	160	692
SAF 10%	14	9	27	158
SAF 50%	63	40	123	742
SAF 60%	76	47	147	888
SAF 10%, Hydrogen 10%	31	19	60	319
SAF 10%, Hydrogen 20%	47	29	90	447
SAF 10%, Hydrogen 30%	62	39	120	570
SAF 10%, Hydrogen 40%	78	49	150	705
SAF 10%, Hydrogen 50%	97	60	189	850
SAF 20%, Hydrogen 10%	43	27	84	465
SAF 20%, Hydrogen 30%	74	47	143	716
SAF 20%, Hydrogen 40%	90	57	175	850
SAF 20%, Hydrogen 50%	108	68	211	997
SAF 30%, Hydrogen 10%	56	35	108	612
SAF 30%, Hydrogen 20%	72	45	138	740

(continued)

Table 5.8 (continued)

Scenarios	Assessment variables			
	Hydrogen capacity 2050 (Kg per year in billion)	CO_2 emissions from H_2 production 2050 (Million tonnes)	CO_2-eq emissions from H_2 production 2050 (Million tonnes)	Green electricity requirements for green H_2 (TWh per year)
SAF 30%, Hydrogen 40%	103	64	199	997
SAF 30%, Hydrogen 50%	122	76	235	1143
SAF 40%, Hydrogen 10%	68	42	131	758
SAF 40%, Hydrogen 30%	83	52	160	886
SAF 40%, Hydrogen 40%	115	72	221	1144
SAF 40%, Hydrogen 50%	134	84	258	1290
SAF 50%, Hydrogen 10%	80	50	155	905
SAF 50%, Hydrogen 20%	96	60	185	1030
SAF 50%, Hydrogen 30%	111	70	214	1155
SAF 50%, Hydrogen 40%	128	80	247	1290
SAF 50%, Hydrogen 50%	146	93	282	1434
SAF 60%, Hydrogen 10%	92	58	180	1048
SAF 60%, Hydrogen 20%	108	67	210	1176
SAF 60%, Hydrogen 30%	123	77	240	1301

(continued)

Table 5.8 (continued)

Scenarios	Assessment variables Hydrogen capacity 2050 (Kg per year in billion)	CO_2 emissions from H_2 production 2050 (Million tonnes)	CO_2-eq emissions from H_2 production 2050 (Million tonnes)	Green electricity requirements for green H_2 (TWh per year)
SAF 60%, Hydrogen 40%	140	87	270	1435
SAF 60%, Hydrogen 50%	158	100	206	1579
SAF 65%, Hydrogen 20%	114	72	221	1248
SAF 65%, Hydrogen 30%	130	81	250	1373
SAF 65%, Hydrogen 40%	146	91	282	1507
SAF 65%, Hydrogen 50%	165	103	317	1651

5.2.3 Sensitivity Analysis

To study the robustness of the model a numerical sensitivity analysis was used because it is considered the most important for addressing future scenarios in the presence of multiple uncertainties (Sterman 2000). For this, Monte Carlo simulation is used to determine the probability of the occurrence of different simulation results that are caused by different parameter values of the exogenous variables. The Monte Carlo simulation is conducted by carrying out 200 simulation runs in which the parameters of the exogenous variables take different values. For this case study, only univariate Monte Carlo sensitivity simulations are conducted instead of multivariate sensitivity analysis. The main reason for this is that the univariate analysis would help more in finding out the most influential and critical variables that could be potentially responsible for the transition of the system. To study the robustness of the model a numerical sensitivity analysis was used because it is considered the most important for addressing future scenarios in the presence of multiple uncertainties (Sterman 2000). For this, Monte Carlo simulation is used to determine the probability of the occurrence of different simulation results that are caused by different parameter values of the exogenous variables.

Table 5.9 shows the parameters considered for the sensitivity analysis and their specific varied values as input. These particular variables are chosen for the sensitivity analysis as they are the most important ones related to delays in the system, and are arguably highly uncertain and likely to be more influential (Sterman 2000). Figure 5.7 shows the results of the univariate sensitivity analysis of different scenarios; only the most important sensitivity runs for the discussions are showcased instead of all scenarios showcased in Table 5.7. Since there is a widely accepted notion in the industry that long-haul and/or medium-haul flights might be mostly or completely replaced by SAF, assessing the univariate sensitivity analysis of the SAF 90% scenario for long and medium-haul flight segments seems relevant: it was found that the model does not show large sensitivity intervals for the variables growth factor, time to adjust the capacity of hydrogen fuel for PtL SAF and time required to change the green electricity capacity for PtL SAF as compared to the time to adjust the production capacity of SAF.

It can be observed in Fig. 5.7 that the confidence interval for a.) and e.) is larger than the confidence interval of any other variable for the same scenario. The parameter setting of 5 years for the variable time to adjust production capacity SAF slightly drifts in the 75% confidence range from the 50% confidence range of the sensitivity analysis over time.

The same phenomenon is not observed for the other variables, instead, their model setting lies in the middle of the 50% confidence interval range. This suggests that the initial setting for the time to adjust production capacity SAF is too low, i.e., it might take longer than 5 years for the SAF capacities to be adjusted. Implicitly, this means that even out of all the sensitivities that are observed for different variables, the time to adjust production capacity SAF turns out to be the most important lever for the SAF 90% scenarios. The sensitivity of the variables for the hydrogen 20–60% scenarios for the short-haul flight segment is showcased in the lower part of the table. For these scenarios, it can be observed that the sensitivity run of the variable time to adjust hydrogen aircraft manufacturing capacity represented in figures i.), n.) and r.) shows the largest confidence interval of all other variables. Furthermore, the shape and progression of the curve are different for this variable. Specifically, the first sudden increase in the variable until the peak in the year 2041 makes the simulation run fall in the upper 100% confidence interval category in the sensitivity simulation for all scenarios.

For the second part, when the system reaches a certain equilibrium presented by the slow descent after the year 2041, the simulated run falls in the lower 100% of the sensitivity simulation for all scenarios. This suggests that for the first increase in the variable, the value of 4 years for hydrogen aircraft manufacturing delay is too low and the latter part too high.

Table 5.9 Parameters considered for sensitivity analysis

Sensitivity parameters				
Parameter	Unit	Model setting	Sensitivity	Source
Growth rate for population, GDP per capita	Dimensionless	±0%	±5%	(World bank 2023b, a)
Time to adjust Hydrogen aircraft manufacturing capacity	Years	Short-haul: 4 Medium-haul: 5 Long-haul: 6	4–10	(Hoelzen et al. 2022a)
Time to adjust capacity hydrogen fuel	Years	7	5–10	(IEA 2021b)
Time required to change in green electricity capacity	Years	7	4–10	(IEA 2021b) and assumptions
Time to adjust production capacity SAF	Years	5	4–10	(ATAG 2020b; ICCT 2021; Eurocontrol 2022b)
Time to adjust capacity hydrogen fuel for PtL SAF	Years	7	4–10	(IEA 2021b)
Time required to change in green electricity capacity for PtL SAF	Years	7	4–10	(IEA 2021b) and assumptions

In other words, the manufacturing capacity of 4 years for short-haul flights is highly ambitious for the system to cope with the initial increase in orders. However, the same value seems to be less ambitious in the latter part of the increase in orders.

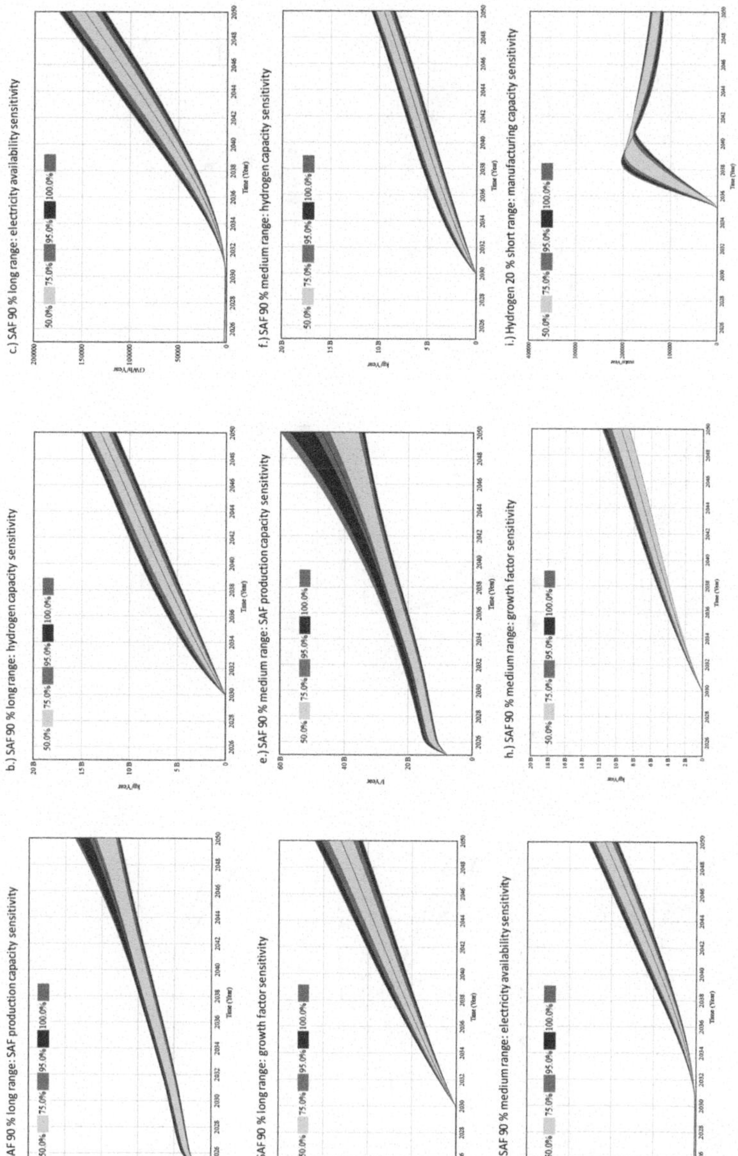

Fig. 5.7 Overview of sensitivity analysis

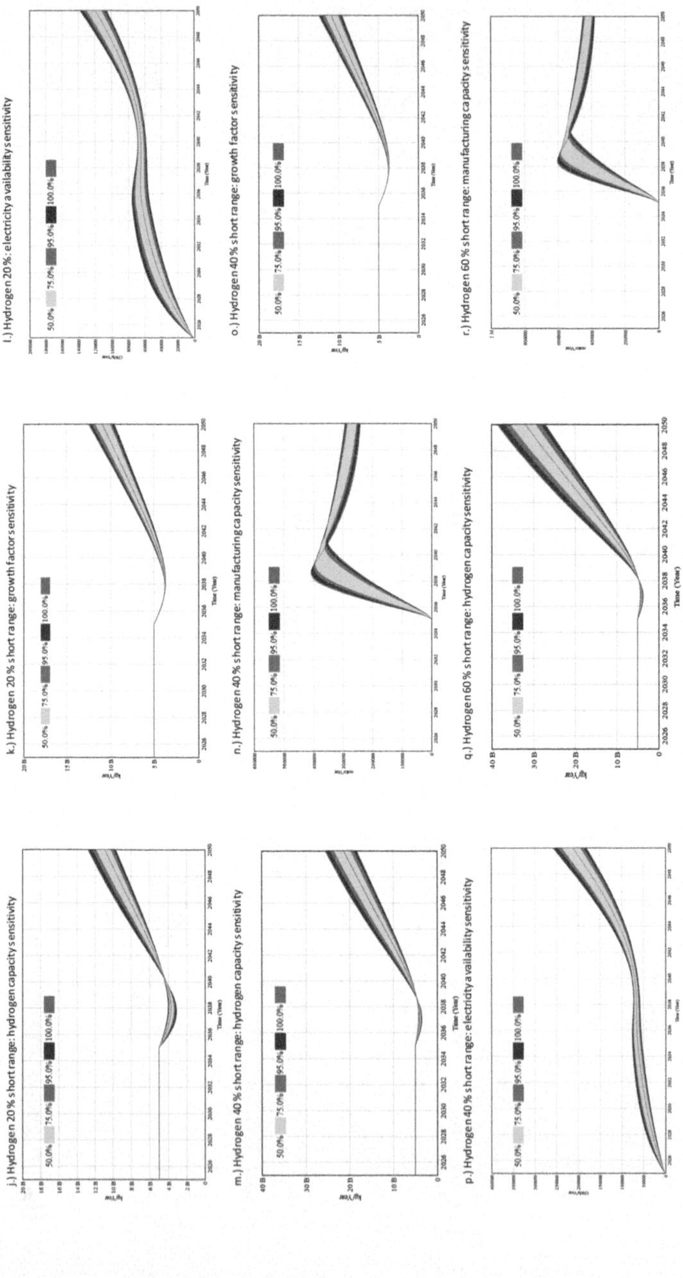

Fig. 5.7 (continued)

Implication of Results and Discussions 6

In this section, the implications of the results presented above are discussed. To this end, this section consists of the explanation of the meaning of the results of the study for different stakeholders, i.e. aircraft manufacturers, fuel producers, regulators and policy-makers, and the airline industry. Along with the discussion of the implications, a few suggestions to tackle the uncertainty regarding the future of the system stakeholders are provided as well.

6.1 Aircraft Manufacturers

From the discussions in the results section, it has been found that the time delay for change in the manufacturing capacity of hydrogen aircraft was found to be the most important parameter for aircraft manufacturers. It was clear in the sensitivity analysis that this variable can play a crucial role in the adoption of hydrogen aircraft especially in the short-haul flight segment. The aircraft manufacturers should pay close attention to the delivery delays associated with the production of a new technology like hydrogen aircraft. Specifically, in the short-haul flight segment, it was found in the results that a value of 4 years' average delay per seat of aircraft delivered to the airlines could be difficult to achieve. This would affect the adoption of the aircraft in the system and cause cascading effects on other stakeholders. Therefore, as mentioned in the study conducted by (Hoelzen et al. 2022a), a delay of 4 years to deliver a short-haul hydrogen aircraft would be extremely difficult. One of the main implications of this study for aircraft manufacturers is that they should focus on closely collaborating with

C. Talwar, *Transition Pathways towards a Low Emission Air Transport System of the Future – A System Level Analysis*, Produktion und Logistik,
https://doi.org/10.1007/978-3-658-46507-0_6

airlines and fuel producers to communicate the real time manufacturing delays of hydrogen aircraft. This close collaboration is important because even if hydrogen fuel capacities are available for the short-haul segment, delays in the delivery of hydrogen aircraft could mean inefficiencies for the system. Furthermore, this would help the fuel producers to adjust their capacities accordingly in an uncertain environment. Since, hydrogen might be a scarce resource in the near future as other sectors such as steel, power and mobility require it as well, this sort of practice would promote efficiencies and correct allocation of resources.

6.2 Fuel Producers

From the results, many implications can be drawn for the aircraft fuel producers. To achieve sustainability in the system, one of the most important aspects for aircraft fuel producers was found to be the delay in increasing the production capacity for SAF and hydrogen. Also, to reach the climate goals, the results showed that an increase in hydrogen and SAF production along with green electricity requirement that is needed in most overall level scenarios is highly ambitious.

In the case of the hydrogen 20% share scenario with hydrogen fuel production delays set at 7 years, the expansion of the hydrogen production required within the first 10 years of introduction is more than three times the initial availability. Furthermore, for the system to be able to support the transition until 2050, more than six times the amount of hydrogen as compared to the initial availability would be required. In this case, the electricity requirement to cater to 30% green hydrogen (also the case for all scenarios discussed in this section) would be 150 times the initial availability of 1850 GWh to almost 290 TWh per year. While this might seem plausible to some extent in terms of hydrogen scaling if not electricity requirements, the increase in production required is five, seven and ten times the initial value for the hydrogen 30%, 40% and 60% scenarios by 2045, respectively. In terms of electricity requirement for 30% green hydrogen, it would need to be increased by roughly 220, 295 and 430 times the initial value for the hydrogen 30%, 40% and 60% scenarios, respectively by 2050.

For SAF, in the case of the SAF 20% scenario with a production delay set to 5 years, the system would require a gain of more than nine times in the SAF production capacity by 2035. Furthermore, this means an eleven, fourteen and seventeen times increase would be required by 2040, 2045 and 2050 respectively for the same scenario. In terms of electricity requirements to produce green hydrogen for 62% of fuel by PtL type of SAF, 164 times increase from the initial capacity of 1850 GWh to 305 TWh by 2050 would be required. This value rises dramatically for the SAF 65% scenario, as the SAF production would need to be ramped by almost 30 times the initial capabilities by 2035. Furthermore, to produce green hydrogen, an increase in green electricity equivalent to 72 times the initial value would be required for 62% of SAF produced through Ptl type SAF. This increase could be unrealistic to achieve given that the level of development of SAF in the last 10 years has been slow. From the above discussion, it can be inferred that an overall system transition could be hard to achieve given the amount of expansion in SAF, hydrogen production and green electricity would be required. While aiming at lower levels of share for hydrogen and SAF might be plausible, these settings would not allow the system to reach the ambitious goal of net zero emissions by IATA for the aviation industry. Achieving a higher overall percentage share of SAF in the system would be hard especially as sensitivity analysis showed that the parameter setting of 5 years of production delays for SAF is too fast. In other words, the sensitivity analysis showed that an increase in capacity would take longer than 5 years to adjust for the changes in production.

If the combination of strategies by the regulators and decision-makers is utilized to achieve the net zero goal for the aviation sector by 2050, then the results show that the requirements for the fuel producers could be achievable. Specifically, in the SAF 90% for the long-haul and hydrogen 20% for the short-haul scenario, the increase in SAF production would be roughly four times greater than the initial value of 7.5 billion liters in 2035. The increase in SAF production for the further period would amount to five, seven and eight times the initial value by 2040, 2045 and 2050, respectively. The total increase in electricity requirements in the case would be 145 times the initial values of 1850 GWh by 2050. It can be observed that the above values are significantly lower than any overall level scenario for the SAF. Furthermore, the hydrogen for the Ptl type of SAF amounts to only 13 billion kg and requiring 150 TWh of green electricity by 2050 for 30% green hydrogen. In the same scenario, aiming for a 20% hydrogen aircraft share in the short-haul flight segment would require the hydrogen capacity to be ramped up by roughly two times by 2050, compared to the initial availability of 5 billion kg and an electricity requirement of 65 times the initial capacity equivalent to 119 TWh. This value is significantly lower than for the hydrogen 20% overall scenario, where three times the hydrogen capacity was needed by 2045 and 150 times the initial electricity capacity increase was required by 2050. By increasing the percentage share of hydrogen aircraft by 30%, 40% and 60% in the short-haul flight segment, it is observed that the increase in production amounts to three, four and six times the initial capacity by 2050, respectively. In this scenario, an increase in electricity requirement was equivalent to 90, 115 and 165 times the initial capacity.

Lastly, the scenario in which all short-haul flights are eliminated, and the medium and long-haul flight segments are powered by 80% or 90% SAF. Given that the overall volume share of these flights in the entire traffic is lower, it appears highly likely that the volumes of SAFs required to meet these requirements are realistic. Specifically, for the 90% SAF in medium and long-haul scenarios, the requirements for long-haul flights remain similar to what was discussed above, but for the medium-haul the required increase in SAF production would be roughly three times the initial value in 2035. This value would increase to four, five and six times the initial capacity by 2040, 2045 and 2050, respectively.

On an overall level, the fuel producers can play a very important role in helping to achieve the transition towards sustainability in the air transport system. One of the most important goals for fuel producers would be to aim for realistic rather than highly ambitious increases in fuel production capacities. Specifically, the hydrogen and SAF fuel producers should aim to provide for an increase in demand in the long-haul and medium-haul flight segments if the system moves more towards sustainability. To achieve this, the regulators, fuel producers and the airlines would have to collaborate to set more realistic goals for the aviation industry rather than the current ambitious ones. Even to achieve realistic goals for fuel producers, one of the biggest challenges could be to source clean energy in the form of green electricity. In this regard, the fuel producers would have to work in close collaborations with energy providers to set targets for future requirements because like hydrogen, green electricity can be a scarce resource in the near future.

6.3 Regulators, Policy and Decision-Makers

Regulators policy and decision-makers have to play one of the most important and decisive roles to help the air transport system achieve low emissions goals. There are many implications and suggestions from the results of this analysis that can help provide clarity to the decision-makers. According to the findings of this study, achieving the IATA net zero 2050 goals for the aviation sector is contingent upon specific conditions. To help achieve this goal, the regulators will have to focus on SAF and hydrogen aircraft technology. The adoption of these two technologies is key to achieving sustainability for the air transport system. Specifically, the results of the analysis showcased that trying to achieve an overall level percentage share of any of the two technologies would not help the system move towards sustainability. One of the main reasons for this is that even if the technology is available the increase in the production of fuels required to meet the adoption demands is highly unrealistic. Furthermore, the amount of increase in electricity requirements to produce green hydrogen to meet the overall percentage share of these technologies would be highly unrealistic.

In this regard, the policy-makers and the regulators should focus on transitioning in the different flight range segments. One of the main results of the analysis shows that policy-makers should focus on the transition of the medium and long-haul flight segments with SAF first, as the increase in production capacity requirement for this segment is smaller compared to the overall scenarios. Additionally, results also suggest that a combination of SAF and hydrogen aircraft for the different flight range segments would be advisable. The simulation results of these scenarios indicate that the requirements for hydrogen, SAF, and electricity are not as high as in the overall scenarios.

For SAF, the sensitivity analysis showed that the time delay to expand SAF production is the most important lever for this technology. Therefore, policy-makers should be mindful that the increase in production capacity ramp up could be slower than what is portrayed in our results. For the hydrogen technology, it can be inferred from the discussion that a 20–30% share of hydrogen aircraft in the short-haul flight segment can be realistic. However, aiming for a higher share might be challenging as the hydrogen requirements dramatically rise, mainly because short-haul flights have the highest share in the overall air traffic volume. Since other industries could also require hydrogen, it might be hard to allocate for aviation. Furthermore, in the sensitivity analysis, the time delay for the manufacturing of hydrogen aircraft was found to be the most important lever for this technology. This means that even if hydrogen capacities are available for the short-haul flight segment the delays in delivery of hydrogen aircraft could result in less adoption in the system. The scenario where all the short-haul flights are either banned or replaced by bus, train or roadways was observed to be the best case scenario. In other words, it can be implied that the increase in the capacities for hydrogen and green electricity for short-haul flights is highly unrealistic and aiming to replace them with other means of transport might be a better option. In this way, the net zero emissions goal set by IATA for the aviation industry might be a possibility.

The regulators and policy-makers will have to adopt a different strategy to achieve the Flightpath 2050 CO_2 emission goal set for the EU by ACARE in 2012. To achieve the Flightpath 2050 CO_2 emissions goal using electric aircraft the regulators would have to use policies. Specifically, the result of the analysis in the first case study showed that jet fuel tax and seat tax policy enacted individually in the model are not enough to reach the Flightpath 2050 CO_2 emissions goal. On the other hand, the fleet restriction policy and a high electricity subsidy were able to show CO_2 emission reduction potential to reach the goals. The results also showed that a strong set of combinations of policy measures was required to reach the emissions reduction goal. In this thesis, two types of combinations were discussed in more detail. The first type is aimed at reducing conventional fleet by obligating the airlines to operate electric aircraft regardless of the costs, i.e., a fleet restriction policy with a seat tax. The second combination focused on encouraging electric and discouraging conventional aircraft at the same time, i.e., a fleet restriction policy with an electricity subsidy, and kerosene tax with electricity subsidy. The results show that both types of combination lead to a higher level of electric aircraft adoption when the strength of the policy measure is high (i.e., high subsidy, high seat tax or 20% fleet restriction). In the multivariate sensitivity analysis, it was observed that the second type shows more volatility in the confidence interval of the CO_2 emissions.

The subsequent univariate sensitivity analysis indicates that the main reason for the volatility is the uncertainty in parameters related to cost of electric aircraft. For the second type of combination, due to the presence of electricity subsidy, changes in the cost parameters cause volatility in electric aircraft adoption and thereby higher volatility in the confidence interval of CO_2 emissions. For the first type of combination, the uncertainty in electric aircraft development parameters does not play an important role, as the airlines have to adopt the new technology regardless of the costs. Therefore, the CO_2 confidence interval does not showcase high volatility.

Overall, the results from the study imply that a policy combination is a better option than enacting individual policies to reach the Flightpath 2050 CO_2 emissions goal if a short-haul electric aircraft is introduced in the market. However, it must be noted that due to the higher costs to use electric aircraft corresponding to parameters used in Table 5.1, an extremely high value of policy parameters was required to achieve any effect in adoption. In the case of the seat tax, it was observed that only a high seat tax equivalent to an increase of 30% of the average unit cost for conventional aircraft was able to make the electric aircraft cost competitive and showcased higher adoption rates. Furthermore, a 75% reduction in the average electricity price in the form of an electricity subsidy was required for the considered period to have 28% electric fleet in the system.

In this regard, policy-makers should keep in mind that if the development parameters for electric aircraft are similar to what is considered in Table 5.1, then implementing such high levels of taxes and subsidies could be politically and economically infeasible, making electric aircraft a highly unattractive option. This further indicates that electric aircraft could be replaced by other technologies, i.e., hydrogen, hydrogen-electric or SAF. Furthermore, policy-makers should also make efforts to monitor the growth rate factors (GDP per capita and population levels) and electricity price as these macro indicators can affect the overall air travel demand and cost attractiveness of operating electric aircraft. Additionally, policy-makers would also need to assess the uncertainties in implementing the policies tested in this study coherently across the European Union states. This is because achieving this kind of coherent agreement across the states could prove to be a huge political challenge.

6.4 Airline Industry

The airline industry is the major stakeholder in the air transport system as it operates aircraft in different flight segments. The airline industry is mainly responsible for the GHG emissions from the use phase. Due to this, the initiatives taken by the airline industry in the near future can determine the outcome of the system. From the results, it can be inferred that for the system to be able to reach the climate goals whether it is IATA net zero 2050 or Flightpath 2050 CO_2 emissions goal, the adoption of new technologies by airlines is imperative. Specifically, to achieve the IATA net zero 2050 emissions goal, the airline industry should focus on medium and long-haul flights using SAF. As discussed in the sections above, the main reason for this implication is that the increased demand for fuel as a result of an increase in the adoption of SAF for the long-haul flight segment can be fulfilled realistically rather than in the overall scenarios. In this regard, the airline industry, fuel producers and regulators would have to work in close collaboration to make sure there are no inefficiencies in the system. Inefficiencies can lead to a situation where there is overproduction of fuel that is not adopted by the airlines or under production of fuel due to rapid adoption by the airline industry. To ensure an increase in capacities of SAF and hydrogen fuel, airlines would have to enter into forward agreements where large amounts of capacities are promised to be bought by airlines. Also, in terms of costs, the adoption of SAFs in the medium and long-haul flights would be highly desirable even if the initial cost of SAFs is higher. The main reason for this is that usually in the airline industry the medium and long-haul flights (Schmitt and Gollnick 2016) are the most revenue and profit generating flights due to the business and first class passengers, therefore, increasing the uptake of SAFs could be achieved without hugely affecting the airline industry's profits.

Regarding the adoption of hydrogen and electric aircraft in the short-haul flights segment, the airlines should focus on achieving a lower percentage share of hydrogen aircraft if the technology is available by 2035. Achieving a lower percentage of hydrogen based propulsion concepts in the short-haul flight could help in reducing use phase emissions from this segment. Since this segment is mostly responsible for large amounts of emissions due to shorter travel time, replacing 10–20% of the fleet can help reduce large emissions over a long period of time until 2050. Electric aircraft on the other hand can only be adopted by airlines if there is a subsidy equivalent to a 75% reduction in the average electricity price. Since short-haul flights are heavily dependent on low costs and other cost advantages to achieve break even or profits, the adoption of new technologies in this segment is difficult. Therefore, the adoption of new technologies in this segment will be highly uncertain in the near future. Furthermore, as showcased in the results, one of the most important parameters for aircraft manufacturers is the uncertainty in the time delays, especially in the short-haul flight segment. Therefore, the presence of delivery uncertainties along with the cost of operations do not allow the short-haul flight segment to become a more sustainable means of travel. In this context, in order to contribute strongly to reaching the Flightpath 2050 CO_2 emissions goal and IATA's net zero 2050 goal, the airlines should aim to change their business strategies that focus more on medium and long-haul flight segments.

Conclusions

7

In this thesis, two different case studies were conducted and their conclusions provide valuable insights for the system and its stakeholders. The first case study suggests that most of the individual policies enacted in the model are not enough to increase electric aircraft's adoption rate to reach the Flightpath 2050 CO_2 emissions goal. The fleet restriction policy together with a high electricity subsidy were the only policies that showed potential to reduce CO_2 emissions and achieve the goals. The first case study also suggests that a strong combination of policy measures is required to reach the emissions reduction goal. This thesis discusses two main types of combinations in detail. The first type focuses on reducing conventional fleets by obligating the airlines to operate electric aircraft regardless of the costs, i.e., a fleet restriction policy with a seat tax. The second combination focuses on encouraging electric aircraft and discouraging conventional aircraft at the same time, i.e., a fleet restriction policy with an electricity subsidy, and kerosene tax with electricity subsidy. The results show that both types of combination lead to a higher level of electric aircraft adoption when the strength of the policy measure is high (i.e., high subsidy, high seat tax or 20% fleet restriction). The multivariate sensitivity analysis highlights the second type's volatility in the confidence interval of the CO_2.

Upon conducting the univariate sensitivity analysis, it was found that the main reason for the volatility is the uncertainty in cost parameters related to electric aircraft development. For the second type of combination, due to the presence of electricity subsidy, changes in the cost parameters cause volatility in electric

C. Talwar, *Transition Pathways towards a Low Emission Air Transport System of the Future – A System Level Analysis*, Produktion und Logistik,
https://doi.org/10.1007/978-3-658-46507-0_7

aircraft adoption because presence of policy can make aircraft cheaper depending upon the value cost parameter takes. Due to this, higher volatility in the confidence interval of CO_2 emissions is observed.

Regarding the conclusions for the second case study, the analysis suggests that aiming for a higher percentage share of more than 20% for either SAF or hydrogen propulsion technologies in the system at an overall level is highly ambitious or unrealistic. The main reason for this is that the capacity increase in the SAF or hydrogen fuel requirement is more than five times the initial capacity for most of the overall share scenarios by 2045. Achieving this large amount of increase in production in the system is highly ambitious. The results suggest that aiming for a lower percentage share of hydrogen in the is a plausible option as the capacity increase is not that unrealistic. However, the ambitious goal of achieving net zero emissions as set by IATA for the aviation industry cannot be reached using this approach. These results are contrary to what is found in the Eurocontrol study that suggested SAF can contribute 41% to the overall emission reduction to reach the net zero goal for aviation. Instead, the results show that aiming for different technologies in different flight segments demonstrates to be a better strategy. The main reason for contrasting results is because the Eurocontrol study did not consider the production level resource requirement for large scale SAF production i.e., green hydrogen and green electricity for PtL type of SAF. It is only after the consideration of the resource requirements it is evident that 41% overall emission reduction using SAFs is not plausible for the system due to the unrealistic increase in production requirements of SAF, green hydrogen and green electricity. Instead, it is found that aiming for more than 80% share of SAF in the long-haul flight segment and a 20% or lower share of hydrogen electric propulsion systems in the short-haul flight segment is a realistic option. In simple words, aiming to decarbonize the long-haul flight segment first and aiming for transition in the short-haul flights later is a better option for the system in the long run.

The conclusions of the thesis provide multiple suggestions to the stakeholders of the system. For policy-makers in the EU regarding reaching the Flightpath 2050 CO_2 emissions goals using policies and electric aircraft, the results imply that a policy combination is a better option than enacting individual policies. However, it must be noted that due to the higher costs of using electric aircraft corresponding to parameters used in Table 5.1, an extremely high value of policy parameters is required to achieve any effect in adoption. It is observed that for electric aircraft to be cost-competitive and have higher adoption rates, a high seat tax equivalent to a 30% increase in the average unit cost for conventional aircraft is necessary. Furthermore, a 75% reduction in the average electricity price in the form of an electricity subsidy is required for the considered period to have 28% electric fleet in the system. In this regard, policy-makers should keep in mind that if the development parameters for electric aircraft are similar to what is considered in Table 5.1, then implementing such high levels of taxes and subsidies can be politically and economically infeasible, making electric aircraft a highly unattractive option. This indicates that electric aircraft might be replaced by other technologies that are more cost attractive and might not require stringent policies, i.e., hydrogen, hydrogen-electric or SAF. Moreover, policy-makers should make efforts to monitor growth rate factors (GDP per capita and population levels) and the electricity price as these macro indicators can affect the overall air travel demand and cost attractiveness of operating an electric aircraft. Additionally, policy-makers need to assess the uncertainties in implementing the policies tested in this study coherently across the European Union states. This is because achieving this kind of coherent agreement across the states can prove to be a huge political challenge.

Apart from the strategies tested for case study two to reach net zero goals by 2050, it is found that in an extreme case scenario where all the short-haul flights are abolished, banned or replaced by bus or train shows to be the easiest method to decarbonize aviation. This outcome suggests that the prospects of significantly expanding hydrogen and green electricity capacities for short-haul flights are challenging, thereby highlighting the viability of transitioning to alternative modes of travel. This sort of mode share change strategy aligns well with the objective of attaining net-zero emissions established by the International Air Transport Association (IATA) for the aviation sector.

There are a few limitations of the work done in this thesis that can be worked on in the near future. The ATS model does not include the airport level operations, e.g., charging time, charging infrastructure and ground operations, or the decision of airlines on aircraft buying and leasing. In the future, incorporating the aforementioned factors in the model can provide insights into airport level CO_2 emissions and reveal the operational difficulties of electric aircraft. Additionally, airport level operations and different methods of green electricity production can also be included in the model to increase the scope and robustness of the model. The model does not consider the varying business strategies used by major stakeholder firms in the system. In the future, an agent based model can be designed and integrated into the ATS system dynamics model to emulate the individual stakeholder firm's different business strategies for new technology adoption. This will provide a better understanding of the set of strategies that are required from the stakeholder firms in the industry to reach the emissions reduction goals. At the same time, it will provide insights into strategies that can prove to be fatal for the system and limit the growth rate of adoption. In this regard, including buying vs. leasing concepts in the model can be an important aspect in showcasing the financial strategies chosen by the different types of airlines in the presence of new technologies. Regarding case study one on Flightpath 2050 CO_2 emissions goal, the focus was on the narrow body aircraft type for short-haul flights under the 1,000 km range and does not take into account regional, long-haul and long-haul feeder flights. The flights offered by other aircraft types such as regional, turbo-prop and wide body aircraft were left out of the analysis. Furthermore, it was assumed that flights offered by each sector of aircraft types contribute equally with respect to the Flightpath 2050 CO_2 emission reduction goals. The influence of willingness to pay by customers for traveling with electric or hydrogen aircraft on air travel demand and the air transport network design is not taken into account but including these aspects in the future works can be a promising extension of the model. Policies that plan to increase the adoption of SAF i.e., blending mandate and energy tax directives are not included in this analysis. In future work, these policies can be incorporated into the ATS system dynamics model to assess their influence on the adoption rates. For future research in this field, the impact of COVID-19 on demand development, airline business operations and its relationship with CO_2 emissions levels can be modeled.

Summary

8

The air transport system is one of the most important industries of the world with its global economic impact estimated at \$2.7 trillion and equivalent to 3.6% of the world's gross domestic product. However, the recent growth of air traffic in the air transport system has led to an increase in concerns related to its impact on the climate. The most harmful greenhouse, gas emissions from the air transport system include CO_2, NO_X, Ozone (O_3), carbon monoxide CO and contrails. Considering the harmful effects of these greenhouse gases, regulators have taken initiatives to reduce emissions from the air transport industry in line with the 1.5°C goal of the Paris Agreement. One of the most crucial goals for the air transport system is achieving the IATA net zero 2050 and Flightpath 2050 CO_2 emissions targets. To accomplish these objectives, it is believed that new propulsion technologies like electric aircraft, hydrogen propulsion aircraft, and sustainable aviation fuels (SAFs) can play an important role. IATA believes that almost 65% of the emissions required to achieve net zero emissions by 2050 can be accomplished through SAFs and 13% of the emissions can be reduced by utilizing new technologies. In light of these developments, there are a lot of uncertainties with regard to how the transition of the system can unfold in the near future as it is hard to conceive the future transition prospects considering the technological developments, policy initiatives and persistent growth in the industry. It can be argued that the decision-makers and policy-makers have laid down ambitious goals for the industry without expounding on the details about production and delivery delays, fuel and green electricity requirements and uncertainties to achieve these goals. Academically, authors in the past have tackled problems related to the aviation industry (Liehr et al. 2001; Sgouridis et al. 2011; Csala

C. Talwar, *Transition Pathways towards a Low Emission Air Transport System of the Future – A System Level Analysis*, Produktion und Logistik,
https://doi.org/10.1007/978-3-658-46507-0_8

and Sgouridis 2013; Pierson and Sterman 2013; Kieckhäfer et al. 2018). However, providing insights with an analysis that considers the sub-systems interaction on a high level along with considerations of new potential technologies and the introduction of policies to assess different sustainability transition pathways has not been conducted in the past. Due to this, clarity regarding the future of aviation in terms of the effects of the adoption of new technologies, introduction of different policies, aircraft manufacturing delays, demand growth patterns and fuel manufacturing delays on the air transport system from a high level dynamic viewpoint does not exist. Furthermore, no study has yet considered flight segment segregation and fuel and electricity requirements together on a system level setting in an attempt to explore the emissions reduction pathways.

Considering the gap in the literature, new technologies and sustainability initiatives, the goal of this thesis was to find out different transition pathways that can help in achieving the sustainable climate goals set by regulators for the air transport industry. Specifically, the study intends to put forth different scenarios in which new technologies can help in achieving the emission reduction goals for different flight segments, taking into consideration the presence of multiple uncertainties and policy initiatives. The secondary goal of this thesis is to highlight the aircraft delivery, fuel production and policy initiative requirements that are essential in achieving the sustainable goals for the aviation industry by 2050. This thesis takes a two-pronged approach to meet its goals. First, this thesis aims to provide industry stakeholders with ways to achieve ambitious goals by analyzing the potential contribution of new technology in different flight ranges. This provides the decision-makers and policy-makers with an initial impression of possibilities on a system level. Second, this thesis intends to fill the gap in the existing literature by developing a model that incorporates sub-systems of the aviation system, such as airline companies, air travel demand, aircraft fuel producers, aircraft manufacturers, environment and aviation regulatory authorities. Developing a system-level model in this way provides the necessary tools for industry stakeholders along with decision-makers to test for future policies and strategies.

To conduct the analysis, a system dynamics methodology was used because it helps to understand complex systems by understanding their nonlinear behavior over time. This methodology has been proven to be highly useful in assessing different strategies and future oriented scenarios based on different parameter values. The system dynamics air transport system (ATS) model was extended by incorporating new technologies, such as electric aircraft, hydrogen propulsion aircraft and SAFs. These new technologies were introduced into the model at different points in time according to the reports and expert opinions in the industry.

To take into account the potential of each technology in helping to reduce emissions in different flight range segments, the model was extended by including the same, i.e. short, medium and long-haul flight segments. The model was further extended by adding a fuel producer segment which included the overall SAF production capacity, hydrogen production capacity, and green electricity capacities to produce green hydrogen. Lastly, the model was extended by adding mechanisms for testing different policies, such as jet fuel tax, seat tax, electricity subsidy and fleet restriction policy. After extensions, the model was validated using the best system dynamics validation techniques and was found to be valid.

To answer the research question, the extended model was applied to two case studies. The first case study investigates whether Flightpath 2050 CO_2 emissions goals can be reached using electric aircraft in the presence of discouragement and encouragement policies, such as jet fuel, tax, electricity subsidy, seat tax policy and fleet restriction policy. The outcome of the first case study intends to assist EU policy-makers in making informed decisions regarding the use of electric aircraft for short-haul flights that are under 1000 km range. The primary objective of the second case study is to analyze the various diffusion strategies for new aircraft propulsion technologies namely SAF, hydrogen electric and hydrogen combustion in three different flight range segments. This was done by generating different scenarios in the presence of uncertain factors such as delays in hydrogen aircraft and fuel manufacturing, SAF manufacturing delays, macroeconomic fluctuations, shifts in flight segment distributions, scale effects resulting from production increases and variations in electricity availability.

The results of the first case study suggest that in most cases enacting jet fuel tax, seat tax, fleet restriction and electricity policy individually is not enough to reach Flightpath 2050 CO_2 emissions goals using electric aircraft in the short-haul flight segment. In contrast, it is found that a strong set of combination of policies is required to achieve the goal. A strong set of combination here refers to a higher parameter value of jet fuel tax, seat tax, electricity subsidy or fleet restriction percentage is a requirement to observe higher electric aircraft adoption effects. This shows that a combination of policies that encourages the use of electric aircraft through the implementation of electricity subsidies while discouraging the use of conventional aircraft by enacting fleet restriction policies is effective. However, the policy combination shows a high level of uncertainty in the Monte Carlo sensitivity analysis. The results of the second case study indicate that achieving net zero emissions by 2050 for aviation is a highly unrealistic goal. The main reason is that the increase in production capacity requirements for SAFs or hydrogen for hydrogen propulsion concepts is highly unrealistic. Rather, it was found that using specific strategies for different range segments seems to be the best way to lower emissions in the system by 2050. Specifically, aiming for more than an 80% share of SAFs in the fuel mix for long-haul flights and a 20% or lower share of hydrogen based propulsion concepts in the short-haul flight segment seems to be a more realistic strategy. This is because, in these scenarios, the increase in production capacity and other resource requirements are not unattainable.

Overall, there are a number of implications from the analysis conducted in this thesis for the air transport system. For the system to be able to move towards net zero emissions by 2050, the decision-makers need to make use of a combination of strategies and policies. By opting for a combination of strategies for different new potential technologies in different range segments, realistic goals for fuel producers, aircraft manufacturers and airlines can be set. Moreover, in the case of Flightpath 2050 CO_2 emissions goal, the results show that a combination of policies is a more feasible and effective way of achieving emission reductions through technology as compared to strong individual policy mechanisms. The analysis also implies that trying to reduce emissions from the short-haul flight segment is extremely hard. One of the major reasons for this is due to the high volume of flights in this segment, the increase in SAF and hydrogen fuel production and aircraft production requirements are highly unrealistic to achieve. Rather, policy and decision-makers should focus more on reducing emissions from medium and long-haul flights using SAFs as the increase in production requirements for these segments is realistically attainable. In terms of technologies, it is a challenge for electric aircraft to play an integral role in reducing emissions from the short-haul flight range segment unless strong policy initiatives are put in place to cover for higher costs. Similarly, it can be quite challenging for the hydrogen fuel-cell propulsion concept to help in reducing emissions from the short-haul flight segment unless a large amount of hydrogen production capacities are in place. More than hydrogen combustion technology, SAFs can play a vital role in achieving emission reduction in the medium and long-haul flight range segments. To make this a reality, airlines have to incentivize fuel producers through forward contracts. This may initially result in a reduction of profits for airlines in the short term as prices of SAFs are higher. This practice can eventually bring the price down due to an increase in production and economies of scale in the long run. Lastly, for the system to move towards sustainability, strong combined coordination from system stakeholders together is required, as it has been shown through this thesis that the system is highly dynamic and decisions taken by one stakeholder deeply affect other stakeholders in the system.

Bibliography

ABBB (2021) Sustainable aviation fuel. https://aviationbenefits.org/environmental-efficiency/climate-action/sustainable-aviation-fuel. Accessed 3 Feb 2022

Abrantes I, Ferreira AF, Magalhães LB, et al (2024) The impact of revolutionary aircraft designs on global aviation emissions. Renew Energy 223:119937. https://doi.org/10.1016/J.RENENE.2024.119937

Abrantes I, Ferreira AF, Silva A, Costa M (2021) Sustainable aviation fuels and imminent technologies—CO2 emissions evolution towards 2050. J Clean Prod 313:127937. https://doi.org/10.1016/J.JCLEPRO.2021.127937

ACARE (2020) FlightPath 2050 Goals. https://www.acare4europe.org/sria/flightpath-2050-goals. Accessed 29 Sep 2020

Agusdinata B, De Klein W (2002) The dynamics of airline alliances. J Air Transp Manag 8:201–211. https://doi.org/10.1016/S0969-6997(01)00052-7

Airbus (2019) Global market forecast. Airbus Industries, Toulouse, France. https://www.airbus.com/aircraft/market/global-market-forecast.html

Airbus (2021) ZEROe—Hydrogen—Airbus. https://www.airbus.com/innovation/zero-emission/hydrogen/zeroe.html. Accessed 26 Aug 2021

Airbus (2016) Mapping demand global market forecast 2016: Highlights. Toulouse, France. https://www.deluca.aero/wp-content/uploads/2016/11/Airbus-GMF-2016-2035-MappingDemand-full_book.pdf

Airbus (2017) Global Market Forecast 2017–2036. Airbus Industries, Toulouse, France. https://www.deluca.aero/wp-content/uploads/2016/11/Airbus-GMF-2016-2035-MappingDemand-full_book.pdf

Airbus (2018) Global market forecast. Airbus Industries, Toulouse, France. https://www.airbus.com/aircraft/market/global-market-forecast.html

Åkerman J (2005) Sustainable air transport—on track in 2050. Transp Res Part D Transp Environ 10:111–126. https://doi.org/10.1016/J.TRD.2004.11.001

Alonso G, Benito A, Lonza L, Kousoulidou M (2014) Investigations on the distribution of air transport traffic and CO2 emissions within the European Union. J Air Transp Manag 36:85–93. https://doi.org/10.1016/j.jairtraman.2013.12.019

Andreoni APMN (2006) A multivariate ARIMA model to forecast air transport demand. In: ETC Conference Papers 2006. Strasbourg, France

C. Talwar, *Transition Pathways towards a Low Emission Air Transport System of the Future – A System Level Analysis*, Produktion und Logistik,
https://doi.org/10.1007/978-3-658-46507-0

ATAG (2018) Aviation benefits beyond borders. Air Transport Action Group, Geneva, Switzerland. https://aviationbenefits.org/media/166711/abbb18_full-report_web.pdf

ATAG (2020a) Facts and figures. https://www.atag.org/facts-figures.html. Accessed 6 Aug 2020

ATAG (2021) Commitment to fly net zero 2050. Geneva, Switzerland. https://aviationbenefits.org/media/167501/atag-net-zero-2050-declaration.pdf

ATAG (2020b) Waypoint 2050. Geneva, Switzerland. https://aviationbenefits.org/media/167187/w2050_full.pdf

ATAG (2019) Fact sheet: Aviation 2050 Goal and the Paris Agreement. https://aviationbenefits.org/downloads/fact-sheet-4-aviation-2050-and-paris-agreement/

Avinor (2018) Electric Aviation Norway. https://avinor.no/contentassets/c29b7a7ec1164e5d8f7500f8fef810cc/jan-otto-reimers_green-future-as_oslo-22-march-2018.pdf

Barke A, Bley T, Thies C, et al (2022) Are Sustainable Aviation Fuels a Viable Option for Decarbonizing Air Transport in Europe? An Environmental and Economic Sustainability Assessment. Appl Sci 12:597. https://doi.org/10.3390/APP12020597

Baroutaji A, Wilberforce T, Ramadan M, Olabi AG (2019) Comprehensive investigation on hydrogen and fuel cell technology in the aviation and aerospace sectors. Renew Sustain Energy Rev 106:31–40. https://doi.org/10.1016/J.RSER.2019.02.022

Bauen A, Bitossi N, German L, et al (2020) Sustainable Aviation Fuels : Status, challenges and prospects of drop-in liquid fuels, hydrogen and electrification in aviation. Johnson Matthey Technol Rev 64:263–278. https://doi.org/10.1595/205651320X15816756012040

BBC (2021) The hydrogen revolution in the skies—BBC Future. https://www.bbc.com/future/article/20210401-the-worlds-first-commercial-hydrogen-plane. Accessed 9 Mar 2022

Bießlich P, Schröder M, Gollnick V, Lütjens K (2014) A System Dynamics Approach to Airport Modelling. In: AIAA AVIATION 2014 -14th AIAA Aviation Technology, Integration, and Operations Conference. American Institute of Aeronautics and Astronautics Inc., Atlanta, GA

Blalock G, Kadiyali V, Simon DH (2007) The Impact of Post-9/11 Airport Security Measures on the Demand for Air Travel. https://doi.org/101086/519816 50:731–755. https://doi.org/10.1086/519816

Bock L, Burkhardt U (2019) Contrail cirrus radiative forcing for future air traffic. Atmos Chem Phys 19:8163–8174. https://doi.org/10.5194/acp-19-8163-2019

Boeing (2016) Current market outlook 2016–2035. Seattle, WA. https://787updates.newairplane.com/Boeing787Updates/media/Boeing787Updates/cmo_print_2016.pdf

Boeing (2018) Orders and Deliveries for January 2018. http://www.boeing.com/commercial/#/orders-deliveries

Bofinger, Heinrich; Strand J (2013) Calculating the Carbon Footprint from Different Classes of Air Travel. World Bank, Washington, DC. http://hdl.handle.net/10986/15602

Booth A, Sutton A, Clowes M, et al (2021) Systematic approaches to a successful literature review, 9th Edition, Third edition. SAGE Publications

Brons M, Pels E, Nijkamp P, Rietveld P (2002) Price elasticities of demand for passenger air travel: a meta-analysis. J Air Transp Manag 8:165–175. https://doi.org/10.1016/S0969-6997(01)00050-3

Cabrera E, Melo de Sousa JM (2022) Use of Sustainable Fuels in Aviation—A Review. Energies 15:2440. https://doi.org/10.3390/EN15072440

Chapman L (2007) Transport and climate change: a review. J Transp Geogr 15:354–367. https://doi.org/10.1016/j.jtrangeo.2006.11.008

Csala D, Sgouridis S (2013) Modeling Dynamic Transitions in the Global Air Transportation System. In: The 31st International Conference of the System Dynamics Society. Cambridge, MA, USA

Dagens industri (2022) Sverige ska bli ledande inom elflyg—Eneroth tillsätter utredning. https://www.di.se/hallbart-naringsliv/sverige-ska-bli-ledande-inom-elflyg-eneroth-tillsatter-utredning/. Accessed 2 Mar 2022

De Jong S, Antonissen K, Hoefnagels R, et al (2017) Life-cycle analysis of greenhouse gas emissions from renewable jet fuel production. Biotechnol Biofuels 10:1–18. https://doi.org/10.1186/S13068-017-0739-7/TABLES/4

Department of transport (2013) UK Aviation Forecasts. London, UK. https://www.nwleics.gov.uk/files/documents/df_t_aviation_forecasts_january_2013/DfT%20Aviation%20Forecasts%20-%20January%202013.pdf

Dray L, Evans A, Reynolds T, et al (2014) Airline fleet replacement funded by a carbon tax: An integrated assessment. Transp Policy 34:75–84. https://doi.org/10.1016/j.tranpol.2014.02.021

Duffner F, Kronemeyer N, Tübke J, et al (2021) Post-lithium-ion battery cell production and its compatibility with lithium-ion cell production infrastructure. Nat Energy 2021 6:123–134. https://doi.org/10.1038/s41560-020-00748-8

EASA (2019) European Aviation Environmental Report. European Union Aviation Safety Agency, Cologne, Germany. https://www.easa.europa.eu/eaer/system/files/usr_uploaded/219473_EASA_EAER_2019_WEB_HI-RES_190311.pdf

EC europa (2020) What is the share of renewable energy in the EU? https://ec.europa.eu/eurostat/cache/infographs/energy/bloc-4c.html. Accessed 7 Aug 2020

Espinoza A, Bautista S, Narváez PC, et al (2017) Sustainability assessment to support governmental biodiesel policy in Colombia: A system dynamics model. J Clean Prod 141:1145–1163. https://doi.org/10.1016/j.jclepro.2016.09.168

Etminan M, Myhre G, Highwood EJ, Shine KP (2016) Radiative forcing of carbon dioxide, methane, and nitrous oxide: A significant revision of the methane radiative forcing. Geophys Res Lett 43:12,614–12,623. https://doi.org/10.1002/2016GL071930

Eurocontrol (2022a) Aviation Outlook 2050. Brussels, Belgium. https://www.eurocontrol.int/publication/eurocontrol-aviation-outlook-2050

Eurocontrol (2022b) Objective Skygreen 2022–2030. Brussels, Belgium. https://www.eurocontrol.int/publication/objective-skygreen-2022-2030

Eurocontrol (2018) European Aviation in 2040—Challenges of Growth. https://www.eurocontrol.int/publication/challenges-growth-2018

European commission (2021a) Proposal for a regulation of the European parliament and the council on ensuring a level playing field for sustainable air transport. https://eur-lex.europa.eu/legal-content/EN/ALL/?uri=CELEX:52021PC0561. Accessed 4 Feb 2022

European commission (2021b) European comission: Excise Duty on Energy. https://ec.europa.eu/taxation_customs/business/excise-duties-alcohol-tobacco-energy/excise-duties-energy_en. Accessed 4 Aug 2021

European Commission (2017) Annual Analyses of the EU Air Transport Market 2016. Brussels, Belgium. https://ec.europa.eu/transport/sites/transport/files/2016_eu_air_transport_industry_analyses_report.pdf

European Commission (2021) Proposal for a Regulation of the European Parliament and of the Council on ensuring a level playing field for sustainable air transport. Brussels, Belgium. https://eur-lex.europa.eu/legal-content/EN/ALL/?uri=CELEX:52021PC0561

European Commission (2013) Airports as Drivers of Economic Success in Peripheral Regions. Brussels, Belgium. https://www.espon.eu/sites/default/files/attachments/01_ADES_DFR.pdf

Faber J, O'Leary A, Leon M de (2018) Taxing aviation fuels in the EU. CE Delft, Delft, The Netherlands. https://cedelft.eu/publications/taxing-aviation-fuels-in-the-eu/

Falk M, Hagsten E (2019) Short-run impact of the flight departure tax on air travel. Int J Tour Res 21:37–44. https://doi.org/10.1002/JTR.2239

Fly zero (2022) Our vision for zero carbon emissions air travel. Cranfield, UK. https://www.ati.org.uk/wp-content/uploads/2022/03/FZO-ALL-REP-0004-FlyZero-Our-Vision-for-Zero-Carbon-Emission-Air-Travel.pdf

Forrester JW, Senge PM (1980) Tests for building confidence in system dynamics models. In: system dynamics. Amsterdam, The Netherlands, pp 209–228

Fossilfritt Sverige (2021) Aviation industry—Fossil-free Sweden. https://fossilfrittsverige.se/roadmap/flygbranschen/. Accessed 2 Mar 2022

Fuel Cells and Hydrogen Joint Undertaking (2020) Hydrogen-powered aviation: a fact-based study of hydrogen technology, economics, and climate impact by 2050. Publications Office, Belgium. https://data.europa.eu/doi/10.2843/471510

Fukui H, Miyoshi C (2017) The impact of aviation fuel tax on fuel consumption and carbon emissions: The case of the US airline industry. Transp Res Part D Transp Environ 50:234–253. https://doi.org/10.1016/j.trd.2016.10.015

Future Flight (2022) Amprius Raises Energy Density Stakes with 450 Wh/kg Battery . https://www.futureflight.aero/news-article/2022-02-11/amprius-raises-energy-density-stakes-450-whkg-battery. Accessed 31 May 2023

Gallet CA, Doucouliagos H (2014) The income elasticity of air travel: A meta-analysis. Ann Tour Res 49:141–155. https://doi.org/10.1016/J.ANNALS.2014.09.006

Gawron B, Białecki T, Janicka A, Suchocki T (2020) Combustion and Emissions Characteristics of the Turbine Engine Fueled with HEFA Blends from Different Feedstocks. Energies 13:1277. https://doi.org/10.3390/EN13051277

Ghaffarzadegan N, Lyneis J, Richardson GP (2011) How small system dynamics models can help the public policy process. Syst Dyn Rev 27:22–44. https://doi.org/10.1002/sdr.442

Ghosh R, Schilling T, Wicke K (2017) Theoretical framework of systems design for the air transportation system including an inherently quantitative philosophy of scenario development. J Air Transp Manag 58:58–67. https://doi.org/10.1016/J.JAIRTRAMAN.2016.09.007

Gnadt AR (2018) Technical and Environmental Assessment of All-Electric 180-Passenger Commercial Aircraft. Massachusetts Institute of Technology

Gnadt AR, Speth RL, Sabnis JS, Barrett SRH (2019) Technical and environmental assessment of all-electric 180-passenger commercial aircraft. Prog Aerosp Sci 105:1–30. https://doi.org/10.1016/J.PAEROSCI.2018.11.002

Godula-Jopek A, Westenberger A (2016) Hydrogen-fueled aeroplanes. Compend Hydrog Energy 67–85. https://doi.org/10.1016/B978-1-78242-364-5.00004-X

Goetz AR, Graham B (2004) Air transport globalization, liberalization and sustainability: Post-2001 policy dynamics in the United States and Europe. J Transp Geogr 12:265–276. https://doi.org/10.1016/j.jtrangeo.2004.08.007

González R, Hosoda EB (2016) Environmental impact of aircraft emissions and aviation fuel tax in Japan. J Air Transp Manag 57:234–240. https://doi.org/10.1016/j.jairtraman.2016.08.006

Gössling S, Lyle C (2021) Transition policies for climatically sustainable aviation. Transp Rev 41:643–658. https://doi.org/10.1080/01441647.2021.1938284

Graham A (2000) Demand for leisure air travel and limits to growth. J Air Transp Manag 6:109–118. https://doi.org/10.1016/S0969-6997(99)00031-9

Grimme W, Jung M (2018) Towards more sustainability?—The development of aviation emissions from Germany between 1995 and 2016. In: Proceedings of the 22nd Air Transport Research Society World Conference. Air Transport Research Society World Conference (ATRS), Seoul, Korea

Gudmundsson S V., Cattaneo M, Redondi R (2021) Forecasting temporal world recovery in air transport markets in the presence of large economic shocks: The case of COVID-19. J Air Transp Manag 91:102007. https://doi.org/10.1016/j.jairtraman.2020.102007

Hoelzen J, Flohr M, Silberhorn D, et al (2022a) H2-powered aviation at airports—Design and economics of LH2 refueling systems. Energy Convers Manag X 14:100206. https://doi.org/10.1016/j.ecmx.2022.100206

Hoelzen J, Silberhorn D, Zill T, et al (2022b) Hydrogen-powered aviation and its reliance on green hydrogen infrastructure—Review and research gaps. Int J Hydrogen Energy 47:3108–3130. https://doi.org/10.1016/J.IJHYDENE.2021.10.239

Hofer C, Dresner ME, Windle RJ (2010) The environmental effects of airline carbon emissions taxation in the US. Transp Res Part D Transp Environ 15:37–45. https://doi.org/10.1016/J.TRD.2009.07.001

Hornung M, Isikveren AT, Cole M, Sizmann A (2013) Ce-Liner—Case Study for eMobility in Air Transportation. In: 2013 Aviation Technology, Integration, and Operations Conference. American Institute of Aeronautics and Astronautics, Reston, Virginia

Hsu H-W, Binyet E, Chang Y-H, Wang W-C (2022) Energy, economic and environmental (3E) analysis for the renewable jet fuel production process. Sustain Prod Consum 33:146–157. https://doi.org/10.1016/j.spc.2022.06.026

IATA (2022a) Net-Zero Carbon Emissions by 2050. https://www.iata.org/flynetzero/. Accessed 5 Jul 2022

IATA (2022b) Fact Sheet 2—Sustainable Aviation Fuel: Technical Certification. https://www.iata.org/contentassets/d13875e9ed784f75bac90f000760e998/saf-technical-certifications.pdf. Accessed 5 Jul 2022

IATA (2020) IATA Annual review. Amsterdam, The Netherlands. https://www.iata.org/contentassets/c81222d96c9a4e0bb4ff6ced0126f0bb/iata-annual-review-2020.pdf

IATA (2022c) Net zero carbon 2050 resolution—Fact Sheet. https://www.iata.org/en/iata-repository/pressroom/fact-sheets/fact-sheet----iata-net-zero-resolution/

IATA (2023) Liquid hydrogen as a potential lowcarbon fuel for aviation. https://www.iata.org/contentassets/d13875e9ed784f75bac90f000760e998/fact_sheet7-hydrogen-fact-sheet_072020.pdf. Accessed 8 Feb 2023

IATA (2021) Net zero 2050: sustainable aviation fuels. https://www.iata.org/en/iata-repository/pressroom/fact-sheets/fact-sheet---alternative-fuels/. Accessed 5 Jul 2021

ICCT (2021) Assessing the sustainability implications of alternative aviation fuels. Berlin, Germany. https://theicct.org/wp-content/uploads/2021/06/Alt-aviation-fuel-sustainability-mar2021.pdf

IEA (2021a) Net Zero by 2050. Paris, France. https://iea.blob.core.windows.net/assets/deebef5d-0c34-4539-9d0c-10b13d840027/NetZeroby2050-ARoadmapfortheGlobalEnergySector_CORR.pdf

IEA (2021b) Global Hydrogen Review. Paris, France. https://iea.blob.core.windows.net/assets/5bd46d7b-906a-4429-abda-e9c507a62341/GlobalHydrogenReview2021.pdf

IEA (2019) Emissions by sector—Greenhouse Gas Emissions from Energy. https://www.iea.org/reports/greenhouse-gas-emissions-from-energy-overview/emissions-by-sector. Accessed 15 Mar 2022

Intervistas (2007) Estimating Air Travel Demand Elasticities. Vancouver, Canada. https://www.iata.org/en/iata-repository/publications/economic-reports/estimating-air-travel-demand-elasticities---by-intervistas/

IPCC (1999) Aviation and the Global Atmosphere. Geneva, Switzerland. https://www.ipcc.ch/report/aviation-and-the-global-atmosphere-2/

Isikveren A, Schmidt M (2014) Future Transport Aircraft Ultra-Low Emissions Technology Options. GARS Workshop 2014

Janić M (2014) Greening commercial air transportation by using liquid hydrogen (LH2) as a fuel. Int J Hydrogen Energy 39:16426–16441. https://doi.org/10.1016/J.IJHYDENE.2014.08.011

Jin W, Xu L, Yang Z (2009) Modeling a policy making framework for urban sustainability: Incorporating system dynamics into the Ecological Footprint. Ecol Econ 68:2938–2949. https://doi.org/10.1016/j.ecolecon.2009.06.010

Joormann I, Talwar C, Spengler, Thomas S (2024) Shaping the future air transport system: Transition pathways assessment of SAF, hydrogen-electric and hydrogen combustion with a system dynamics simulation model. https://doi.org/10.24355/dbbs.084-202404150946-0

Kadyk T, Winnefeld C, Hanke-Rauschenbach R, Krewer U (2018) Analysis and Design of Fuel Cell Systems for Aviation. Energies 11:375. https://doi.org/10.3390/en11020375

Kakoulaki G, Kougias I, Taylor N, et al (2021) Green hydrogen in Europe—A regional assessment: Substituting existing production with electrolysis powered by renewables. Energy Convers Manag 228:113649. https://doi.org/10.1016/J.ENCONMAN.2020.113649

Kapmeier F, Gonçalves P (2018) Wasted paradise? Policies for Small Island States to manage tourism-driven growth while controlling waste generation: the case of the Maldives. Syst Dyn Rev 34:172–221. https://doi.org/10.1002/sdr.1607

Karpuk S, Elham A (2021) Influence of Novel Airframe Technologies on the Feasibility of Fully-Electric Regional Aviation. Aerosp 2021 8:163. https://doi.org/10.3390/AEROSPACE8060163

Khan MAH, Brierley J, Tait KN, et al (2022) The Emissions of Water Vapour and NOx from Modelled Hydrogen-Fuelled Aircraft and the Impact of NOx Reduction on Climate Compared with Kerosene-Fuelled Aircraft. Atmosphere (Basel) 13:1660. https://doi.org/10.3390/ATMOS13101660

Kieckhäfer K, Quante G, Müller C, et al (2018) Simulation-Based Analysis of the Potential of Alternative Fuels towards Reducing CO_2 Emissions from Aviation. Energies 11:186. https://doi.org/10.3390/en11010186

Kleer B, Cronrath EM, Zock A (2008) Market development of airline companies: A system dynamics view on strategic movements. In: In Proceedings of the 26th International Conference of the System Dynamics Society. Athens, Greece

Kluge U, Paul A, Cook A, Cristóbal S (2017) Factors influencing European passenger demand for air transport. 21st Air Transport Research Society World Conference 2017 (ATRS)

Kopsch F (2012) A demand model for domestic air travel in Sweden. J Air Transp Manag 20:46–48. https://doi.org/10.1016/J.JAIRTRAMAN.2011.11.006

Kousoulidou M, Lonza L (2016) Biofuels in aviation: Fuel demand and CO2 emissions evolution in Europe toward 2030. Transp Res Part D Transp Environ 46:166–181. https://doi.org/10.1016/J.TRD.2016.03.018

Krenek A, Schratzenstaller M (2016) Sustainability-oriented EU Taxes: The Example of a European Carbon-based Flight Ticket Tax. Austrian Institute of Economic Research. https://www.umu.se/globalassets/organisation/fakulteter/samfak/fairtax/fulltext01_930270.pdf

Krenek A, Schratzenstaller M (2017) Sustainability-oriented tax-based own resources for the European Union: a European carbon-based flight ticket tax. Empirica 44:665–686. https://doi.org/10.1007/S10663-017-9381-7/TABLES/4

Larsson J, Elofsson A, Sterner T, Åkerman J (2019) International and national climate policies for aviation: a review. Clim Policy 19:787–799. https://doi.org/10.1080/14693062.2018.1562871

Lee DS, Fahey DW, Forster PM, et al (2009) Aviation and global climate change in the 21st century. Atmos Environ 43:3520–3537. https://doi.org/10.1016/J.ATMOSENV.2009.04.024

Lee DS, Fahey DW, Skowron A, et al (2021) The contribution of global aviation to anthropogenic climate forcing for 2000 to 2018. Atmos Environ 244:117834. https://doi.org/10.1016/J.ATMOSENV.2020.117834

Liehr M, Größler A, Klein M, Milling PM (2001) Cycles in the sky: understanding and managing business cycles in the airline market. Syst Dyn Rev 17:311–332. https://doi.org/10.1002/sdr.226

Lyneis JM (2000) System dynamics for market forecasting and structural analysis. Syst Dyn Rev 16:3–25. https://doi.org/10.1002/(SICI)1099-1727(200021)16:1<3::AID-SDR183>3.0.CO;2-5

Lyneis JM (1999) System dynamics for business strategy: a phased approach. Syst Dyn Rev 15:37–70. https://doi.org/10.1002/(SICI)1099-1727(199921)15:1

Mangold J, Silberhorn D, Moebs N, et al (2022) Refueling of LH2 Aircraft-Assessment of Turnaround Procedures and Aircraft Design Implication. https://doi.org/10.3390/en15072475

Mauler L, Duffner F, Zeier WG, Leker J (2021) Battery cost forecasting: a review of methods and results with an outlook to 2050. Energy Environ Sci 14:4712–4739. https://doi.org/10.1039/D1EE01530C

Mayor K, Tol RSJ (2010) The impact of European climate change regulations on international tourist markets. Transp Res Part D Transp Environ 15:26–36. https://doi.org/10.1016/J.TRD.2009.07.002

Mayor K, Tol RSJ (2007) The impact of the UK aviation tax on carbon dioxide emissions and visitor numbers. Transp Policy 14:507–513. https://doi.org/10.1016/j.tranpol.2007.07.002

Mckinsey (2020) How airlines can chart a path to zero-carbon flying. https://www.mckinsey.com/industries/travel-logistics-and-transport-infrastructure/our-insights/how-airlines-can-chart-a-path-to-zero-carbon-flying. Accessed 6 Aug 2020

Michelmann J, Schmalz U, Becker A, et al (2023) Influence of COVID-19 on air travel—A scenario study toward future trusted aviation. J Air Transp Manag 106:102325. https://doi.org/10.1016/J.JAIRTRAMAN.2022.102325

Möller C, Weiermair K, Wintersberger E (2007) The changing travel behaviour of Austria's ageing population and its impact on tourism. Tour Rev 62:15–20. https://doi.org/10.1108/16605370780000316/FULL/XML

NASA (2008) Safeguarding Our Atmosphere. https://www.nasa.gov/centers/glenn/about/fs10grc.html. Accessed 6 Aug 2020

Ng KS, Farooq D, Yang A (2021) Global biorenewable development strategies for sustainable aviation fuel production. Renew Sustain Energy Rev 150:111502. https://doi.org/10.1016/J.RSER.2021.111502

Nicolay S, Karpuk S, Liu Y, Elham A (2021) Conceptual design and optimization of a general aviation aircraft with fuel cells and hydrogen. Int J Hydrogen Energy 46:32676–32694. https://doi.org/10.1016/J.IJHYDENE.2021.07.127

O'Connell A, Kousoulidou M, Lonza L, Weindorf W (2019) Considerations on GHG emissions and energy balances of promising aviation biofuel pathways. Renew Sustain Energy Rev 101:504–515. https://doi.org/10.1016/J.RSER.2018.11.033

Öko-Institut e.V. (2018) Ensuring a Sustainable Supply of Raw Materials for Electric Vehicles: A Synthesis Paper on Raw Material Needs for Batteries and Fuel Cells. https://www.agora-verkehrswende.de/fileadmin/Projekte/2017/Nachhaltige_Rohstoffversorgung_Elektromobilitaet/Agora_Verkehrswende_Rohstoffestrategien_EN_WEB.pdf.

Pierson K, Sterman JD (2013) Cyclical dynamics of airline industry earnings. Syst Dyn Rev 29:129–156. https://doi.org/10.1002/sdr.1501

Pigou AC (1929) The Economics of Welfare. Macmillan and Company, limited.

Placke T, Kloepsch R, Dühnen S, Winter M (2017) Lithium ion, lithium metal, and alternative rechargeable battery technologies: the odyssey for high energy density. J Solid State Electrochem 21:1939–1964. https://doi.org/10.1007/S10008-017-3610-7

Ploetner KO, Schmidt M, Baranowski D, et al (2013) Operating cost estimation for electric-powered transport aircraft. In: 2013 Aviation Technology, Integration, and Operations Conference

Ponater M, Pechtl S, Sausen R, et al (2006) Potential of the cryoplane technology to reduce aircraft climate impact: A state-of-the-art assessment. Atmos Environ 40:6928–6944. https://doi.org/10.1016/J.ATMOSENV.2006.06.036

Posada J (2017) Advances, Challenges and Future of All Electric Aircraft. Pascual Bravo University Institution, Medellín, Colombia

Qiu R, Xu J, Xie H, et al (2020) Carbon tax incentive policy towards air passenger transport carbon emissions reduction. Transp Res Part D Transp Environ 85:102441. https://doi.org/10.1016/j.trd.2020.102441

Quan C (1997) A Systems Dynamics Model for the Development of China's Aviation System. Airport Modeling and Simulation. Virginia Tech. http://hdl.handle.net/10919/44822

Renewable Energy World (2011) The Silent Energy [R]evolution 20 Years in the Making. https://www.renewableenergyworld.com/2011/08/26/the-silent-energy-revolution-20-years-in-the-making/#gref. Accessed 7 Aug 2020

Roland Berger (2020a) Sustainable aviation fuels key for the future of air travel. London, UK. https://www.rolandberger.com/en/Insights/Publications/Sustainable-aviation-fuels-key-for-the-future-of-air-travel.html

Roland Berger (2020b) Hydrogen: A future fuel for aviation? Munich, Germany. https://www.rolandberger.com/en/Insights/Publications/Hydrogen-A-future-fuel-for-aviation.html

Rozman Č, Pažek K, Prišenk J, et al (2013) System dynamics model for policy scenarios of organic farming development. Organizacija 45:212–218. https://doi.org/10.2478/v10051-012-0016-2

Sahoo S, Zhao X, Kyprianidis K (2020) A Review of Concepts, Benefits, and Challenges for Future Electrical Propulsion-Based Aircraft. Aerospace 7:44. https://doi.org/10.3390/AEROSPACE7040044

Samara E, Georgiadis P, Bakouros I (2012) The impact of innovation policies on the performance of national innovation systems: A system dynamics analysis. Technovation 32:624–638. https://doi.org/10.1016/j.technovation.2012.06.002

Schäfer AW, Barrett SRH, Doyme K, et al (2019) Technological, economic and environmental prospects of all-electric aircraft. Nat Energy 4:160–166. https://doi.org/10.1038/s41560-018-0294-x

Schmidt M, Paul A, Cole M, Plötner KO (2014) Implications of Aircraft Concepts Using Alternative Energy on Ground Operation. In: 18h Air Transport Research Society (ATRS) World Conference. Bordeaux, France

Schmitt D, Gollnick V (2016) Air Transport System. Springer Vienna, Vienna

Seetaram N, Song H, Ye S, Page S (2018) Estimating willingness to pay air passenger duty. Ann Tour Res 72:85–97. https://doi.org/10.1016/J.ANNALS.2018.07.001

Sgouridis S, Bonnefoy PA, Hansman RJ (2011) Air transportation in a carbon constrained world: Long-term dynamics of policies and strategies for mitigating the carbon footprint of commercial aviation. Transp Res Part A Policy Pract 45:1077–1091. https://doi.org/10.1016/j.tra.2010.03.019

Shahriar MF, Khanal A (2022) The current techno-economic, environmental, policy status and perspectives of sustainable aviation fuel (SAF). Fuel 325:124905. https://doi.org/10.1016/J.FUEL.2022.124905

Staack I, Sobron A, Krus P (2021) The potential of full-electric aircraft for civil transportation: from the Breguet range equation to operational aspects. CEAS Aeronaut J 12:803–819. https://doi.org/10.1007/S13272-021-00530-W/FIGURES/11

Staples MD, Malina R, Suresh P, et al (2018) Aviation CO2 emissions reductions from the use of alternative jet fuels. Energy Policy 114:342–354. https://doi.org/10.1016/J.ENPOL.2017.12.007

Sterman JD (2000) Business Dynamics. System Thinking and Modelling for a Complex World. McGraw Hill, Boston, USA

Suryani E, Chou SY, Chen CH (2010) Air passenger demand forecasting and passenger terminal capacity expansion: A system dynamics framework. Expert Syst Appl 37:2324–2339. https://doi.org/10.1016/J.ESWA.2009.07.041

Tabelin CB, Dallas J, Casanova S, et al (2021) Towards a low-carbon society: A review of lithium resource availability, challenges and innovations in mining, extraction and recycling, and future perspectives. Miner Eng 163:106743. https://doi.org/10.1016/J.MINENG.2020.106743

Talwar C, Joormann I, Ginster R (2021a) A System Dynamics Model of the Air Transport System.: https://doi.org/10.24355/DBBS.084-202105100741-0

Talwar C, Joormann I, Ginster R, Spengler TS (2021b) A System Dynamics Model of the Air Transport System. Braunschweig, Germany. https://doi.org/10.24355/DBBS.084-202105101036-0

Talwar C, Joormann I, Ginster R, Spengler TS (2023) How much can electric aircraft contribute to reaching the Flightpath 2050 CO2 emissions goal? A system dynamics approach for european short haul flights. J Air Transp Manag 112:102455. https://doi.org/10.1016/J.JAIRTRAMAN.2023.102455

Tanzil AH, Brandt K, Wolcott M, et al (2021) Strategic assessment of sustainable aviation fuel production technologies: Yield improvement and cost reduction opportunities. Biomass and Bioenergy 145:105942. https://doi.org/10.1016/J.BIOMBIOE.2020.105942

Theil H (1966) Applied Economic Forecasting. North Holland Publishing Company, Amsterdam, The Netherlands

Tillmann AM, Joormann I, Ammann SCL (2023) Reproducible air passenger demand estimation. J Air Transp Manag 112:102462. https://doi.org/10.1016/J.JAIRTRAMAN.2023.102462

Tol RSJ (2007) The impact of a carbon tax on international tourism. Transp Res Part D Transp Environ 12:129–142. https://doi.org/10.1016/J.TRD.2007.01.004

Umweltbundesamt (2018) High costs when environmental protection is neglected. https://www.umweltbundesamt.de/en/press/pressinformation/high-costs-when-environmental-protection-is. Accessed 9 Feb 2021

Umweltbundesamt (2019) Methodenkonvention 3.0 zur Ermittlung von Umweltkosten—Kostensätze. Dessau-Roßlau, Germany. https://www.umweltbundesamt.de/sites/default/files/medien/1410/publikationen/2019-02-11_methodenkonvention-3-0_kostensaetze_korr.pdf

USGS (2021) Sulphur. Virginia, USA. https://pubs.usgs.gov/periodicals/mcs2021/mcs2021-sulfur.pdf

Verstraete D (2013) Long range transport aircraft using hydrogen fuel. Int J Hydrogen Energy 38:14824–14831. https://doi.org/10.1016/J.IJHYDENE.2013.09.021

Vespermann J, Wald A (2011) Much Ado about Nothing?—An analysis of economic impacts and ecologic effects of the EU-emission trading scheme in the aviation industry. Transp Res Part A Policy Pract 45:1066–1076. https://doi.org/10.1016/j.tra.2010.03.005

Von Drachenfels N, Engels P, Husmann J, et al (2021) Scale-Up of Pilot Line Battery Cell Manufacturing Life Cycle Inventory Models for Life Cycle Assessment. Procedia CIRP 98:13–18. https://doi.org/10.1016/J.PROCIR.2020.12.002

Winnefeld C, Kadyk T, Bensmann B, et al (2018) Modelling and Designing Cryogenic Hydrogen Tanks for Future Aircraft Applications. Energies 11:105. https://doi.org/10.3390/EN11010105

World bank (2023a) Population. https://data.worldbank.org/indicator/SP.POP.TOTL. Accessed 24 Mar 2023

World bank (2023b) GDP per capita (constant 2015 US$). https://data.worldbank.org/indicator/NY.GDP.PCAP.KD. Accessed 24 Mar 2023

World Bank (2021) GDP per capita (constant 2015 US$)—European Union. https://data.worldbank.org/indicator/NY.GDP.PCAP.KD?end=2018&locations=EU&start=1990&view=chart. Accessed 1 Mar 2024

World Economic forum (2021) Guidelines for a Sustainable Aviation Fuel Blending Mandate in Europe. Geneva, Switzerland. https://www3.weforum.org/docs/WEF_CST_EU_Policy_2021.pdf

World Economic Forum (2020) Clean Skies for Tomorrow: Sustainable Aviation Fuels as a Pathway to Net-Zero Aviation. Geneva, Switzerland. https://www3.weforum.org/docs/WEF_Clean_Skies_Tomorrow_SAF_Analytics_2020.pdf

World Tourism Organization (UNWTO) (2016) Affiliate Members Global Reports—Volume thirteen—The Power of Youth Travel. Madrid, Spain. https://www.e-unwto.org/doi/epdf/10.18111/9789284417162

Yilmaz I, Ilbaş M, Taştan M, Tarhan C (2012) Investigation of hydrogen usage in aviation industry. Energy Convers Manag 63:63–69. https://doi.org/10.1016/J.ENCONMAN.2011.12.032

yle (2022) Finland unveils plan for electric aircraft routes across Lapland region. https://yle.fi/news/3-12302109. Accessed 2 Mar 2022

Zhang L, Butler TL, Yang, B (2020) Recent Trends, Opportunities and Challenges of Sustainable Aviation Fuel. In: Green Energy to Sustainability. Wiley, pp 85–110

The manufacturer's authorised representative in the EU is Springer Nature Customer Service Centre GmbH, Europaplatz 3, 69115 Heidelberg, Germany. If you have any concerns regarding our products, please contact ProductSafety@springernature.com

Printed and bound by CPI Group (UK) Ltd, Croydon, CR0 4YY
01/10/2026
02225951-0002